THE NUCLEAR ENVELOPE AND THE NUCLEAR MATRIX

THE WISTAR SYMPOSIUM SERIES

Volume 1

INTRODUCTION OF MACROMOLECULES INTO VIABLE MAMMALIAN CELLS
Renato Baserga, Carlo Croce, and Giovanni Rovera, *Editors*

Volume 2

THE NUCLEAR ENVELOPE AND THE NUCLEAR MATRIX
Gerd G. Maul, *Editor*

THE NUCLEAR ENVELOPE AND THE NUCLEAR MATRIX

PROCEEDINGS OF THE SECOND WISTAR SYMPOSIUM, HELD AT
SUGARLOAF CONFERENCE CENTER
PHILADELPHIA, PENNSYLVANIA
SEPTEMBER 27–29, 1981

Editor

GERD G. MAUL
The Wistar Institute
Philadelphia, Pennsylvania

Alan R. Liss, Inc. New York

Address all Inquiries to the Publisher
Alan R. Liss, Inc., 150 Fifth Avenue, New York, NY 10011

Copyright © 1982 Alan R. Liss, Inc.

Printed in the United States of America.

Under the conditions stated below the owner of copyright for this book hereby grants permission to users to make photocopy reproductions of any part or all of its contents for personal or internal organizational use, or for personal or internal use of specific clients. This consent is given on the condition that the copier pay the stated per-copy fee through the Copyright Clearance Center, Incorporated, 21 Congress Street, Salem, MA 01970, as listed in the most current issue of "Permissions to Photocopy" (Publisher's Fee List, distributed by CCC, Inc.), for copying beyond that permitted by sections 107 or 108 of the US Copyright Law. This consent does not extend to other kinds of copying, such as copying for general distribution, for advertising or promotional purposes, for creating new collective works, or for resale.

Library of Congress Cataloging in Publication Data

Main entry under title:

The Nuclear envelope and the nuclear matrix.
   (The Wistar symposium series ; v. 2)
   Includes bibliographies and index.
   1. Nuclear membranes—Congresses. 2. Cell nuclei—Congresses. I. Maul, Gerd G. II. Wistar Institute of Anatomy and Biology. III. Series.
QH601.2.N83   574.87'32   82-15259
ISBN 0-8451-2001-8   AACR2

Contents

List of Contributors and Participants	vii
Preface Gerd G. Maul	ix
Aspects of a Hypothetical Nucleocytoplasmic Transport Mechanism Gerd G. Maul	1
Attempted Isolation of the Nuclear Pore Complex: Evidence for a Polydisperse Complex of Polypeptides Robert P. Aaronson, Laura A. Coruzzi, and Kenneth Schmidt	13
Protein Kinase and Phosphatase Reactions of the Nuclear Envelope Khalil Ahmed and Randolph C. Steer	31
Phospholipid Metabolism With Respect to Nuclear Envelope Reformation in Late Mitosis in HeLa S_3 Cells S.M. Henry and L.D. Hodge	47
The Uptake of Nuclear Proteins in Oocytes Carl M. Feldherr and Joseph A. Ogburn	63
Mechanisms of Nuclear Protein Concentration Philip L. Paine	75
RNA Transport Studies in a Cell-Free System Dorothy E. Schumm	85
Comparison of Methods for Studying RNA Efflux From Isolated Nuclei Paul S. Agutter	91
Effect of Tryptophan on Nuclear Envelope in Rat Liver: Evidence for Increased Nuclear RNA Release Challakonda N. Murty, Ethel Verney, and Herschel Sidransky	111
Immunological Probes to Investigate the Role of the Nuclear Envelope in RNA Transport F.A. Baglia and G.G. Maul	129
The Major Karyoskeletal Proteins of Oocytes and Erythrocytes of Xenopus Laevis Georg Krohne, Reimer Stick, Peter Hausen, Jurgen A. Kleinschmidt, Marie-Christine Dabauvalle, and Werner W. Franke	135

Characterization of an Isolated Nuclear Matrix Fraction Obtained From Embryos of Drosophila Melanogaster
Paul A. Fisher, Miguel Berrios and Gunter Blobel 145

The Nuclear Lamins: Their Properties and Functions
Keith R. Shelton, Patsy M. Egle, and David L. Cochran 157

The Further Evidence for Physiological Association of Newly Replicated DNA With the Nuclear Matrix
Bert Vogelstein, Barry Nelkin, Drew Pardoll, and Brett F. Hunt 169

Nuclear Matrix Organization and DNA Replication
Ronald Berezney, Joseph Basler, Linda A. Buchholtz, Harold C. Smith, and Alan J. Siegel 183

The Attachment of Replicating DNA to the Nuclear Matrix
Friedrich Wanka, Anna C.M. Pieck, Ad G. M. Bekers, and Leon H.F. Mullenders 199

The Role of the Nuclear Matrix in Polyoma DNA Replication
Vincent Pigiet and Glen W. Humphrey 213

Crosslinking Experiments in Nuclear Matrix: Nonhistone Proteins to Histones and SnRNA to HnRNA
A. Oscar Pogo, Luis Cornudella, Alice E. Grebanier, Roman Procyck, and Valerie Zbrzezna 223

On the Binding of Host and Viral RNA to a Nuclear Matrix
Walther J. van Venrooij, Chris A.G. van Eekelen, Edwin C.M. Mariman, and Rita J. Reinders 235

The Nuclear Matrix in Steroid Hormone Action
Evelyn R. Barrack 247

Heterogeneity of Estrogen-Binding Sites And the Nuclear Matrix
J.H. Clark and B.M. Markaverich 259

A Model for Nucleocytoplasmic Transport of Ribonucleoprotein Particles
Gary A. Clawson and Edward A. Smuckler 271

The Interaction of Nuclear Reactant Drugs With the Nuclear Membrane and Nuclear Matrix
Kenneth D. Tew 279

The Topology of DNA Loops: A Possible Link Between the Nuclear Matrix Structure and Nucleic Acid Function
Andrew P. Feinberg and Donald S. Coffey 293

Evolution of the Nuclear Matrix and Envelope
T. Cavalier-Smith 307

Index 319

Contributors and Participants

Robert P. Aaronson, Mt. Sinai School of Medicine, New York, New York [13]*
Paul S. Agutter, Napier College, Edinburgh, United Kingdom [91]*
Khalil Ahmed, University of Minnesota, Minneapolis, Minnesota [31]*
F.A. Baglia, The Wistar Institute of Anatomy and Biology, Philadelphia, Pennsylvania [129]*
Evelyn R. Barrack, Johns Hopkins University School of Medicine, Baltimore, Maryland [247]*
Joseph Basler, State University of New York, Buffalo, New York [183]
Ad G.M. Bekers, University of Nijmegen, Nijmegen, The Netherlands [199]
Ronald Berezney, State University of New York, Buffalo, New York [183]
Miguel Berrios, Rockefeller University, New York, New York [145]
Gunter Blobel, Rockefeller University, New York, New York [145]
Linda A. Buchholtz, State University of New York, Buffalo, New York [183]
T. Cavalier-Smith, King's College, London, England [307]*
J.H. Clark, Baylor College of Medicine, Houston, Texas [259]*
Gary A. Clawson, University of California School of Medicine, San Francisco, California [271]
David L. Cochran, Medical College of Virginia, Virginia Commonwealth University, Richmond, Virginia [157]
Donald S. Coffey, Johns Hopkins University School of Medicine, Baltimore, Maryland [293]*
Luis Cornudella, The Lindsley F. Kimball Research Institute of the New York Blood Center, New York, New York [223]
Laura A. Coruzzi, Mt. Sinai School of Medicine, New York, New York [13]
Marie-Christine Dabauvalle, German Cancer Center, Heidelberg, Federal Republic of Germany [135]
Patsy M. Egle, Medical College of Virginia, Virginia Commonwealth University, Richmond, Virginia [157]
Andrew P. Feinberg, Johns Hopkins University School of Medicine, Baltimore, Maryland [293]
Carl M. Feldherr, University of Florida College of Medicine, Gainesville, Florida [63]*
Paul A. Fisher, Rockefeller University, New York, New York [145]*
Werner W. Franke, German Cancer Center, Heidelberg, Federal Republic of Germany [135]
Alice E. Grebanier, The Lindsley F. Kimball Research Institute of the New York Blood Center, New York, New York [223]
Peter Hausen, Max-Planck-Institut für Virusforschung, Tubingen, Federal Republic of Germany [135]
S.M. Henry, Medical College of Georgia, Augusta, Georgia [47]
L.D. Hodge, Medical College of Georgia, Augusta, Georgia [47]*
Glen W. Humphrey, Johns Hopkins University, Baltimore, Maryland [213]
Brett F. Hunt, Johns Hopkins University School of Medicine, Baltimore, Maryland [169]

The number in bold type following the contributor's affiliation is the first page of that contributor's chapter.

*Denotes a contributor who was in attendance at the symposium.

Jurgen A. Kleinschmidt, German Cancer Center, Heidelberg, Federal Republic of Germany [135]
Georg Krohne, German Cancer Center, Heidelberg, Federal Republic of Germany [135]*
Edwin C.M. Mariman, University of Nijmegen, The Netherlands [235]
B.M. Markaverich, Baylor College of Medicine, Houston, Texas [259]
Gerd G. Maul, The Wistar Institute, Philadelphia, Pennsylvania [ix, 1, 129]*
Leon H.F. Mullenders, University of Nijmegen, Nijmegen, The Netherlands [199]
Challakonda N. Murty, George Washington University Medical Center, Washington, D.C. [111]*
Barry Nelkin, Johns Hopkins University School of Medicine, Baltimore, Maryland [169]
Joseph A. Ogburn, University of Florida College of Medicine, Gainesville, Florida [63]
Philip L. Paine, Michigan Cancer Foundation, Detroit, Michigan [75]*
Drew Pardoll, Johns Hopkins University School of Medicine, Baltimore, Maryland [169]
Anna C.M. Pieck, University of Nijmegen, Nijmegen, The Netherlands [199]
Vincent Pigiet, Johns Hopkins University, Baltimore, Maryland [213]*
A. Oscar Pogo, The Lindsley F. Kimball Research Institute of the New York Blood Center, New York, New York [223]*
Roman Procyck, The Lindsley F. Kimball Research Institute of the New York Blood Center, New York, New York [223]
Rita J. Reinders, University of Nijmegen, The Netherlands [235]
Kenneth Schmidt, Mt. Sinai School of Medicine, New York, New York [13]
Dorothy E. Schumm, Ohio State University College of Medicine, Columbus, Ohio [85]*
Keith R. Shelton, Medical College of Virginia, Virginia Commonwealth University, Richmond, Virginia [157]*
Herschel Sidransky, George Washington University Medical Center, Washington, D.C. [111]
Alan J. Siegel, State University of New York, Buffalo, New York [183]
Harold C. Smith, State University of New York, Buffalo, New York [183]
Edward A. Smuckler, University of California School of Medicine, San Francisco, California [271]*
Randolph C. Steer, University of Minnesota, Minneapolis, Minnesota [31]
Reimer Stick, Max-Planck-Institut für Virusforschung, Tubingen, Federal Republic of Germany [135]
Kenneth D. Tew, Lombardi Cancer Center, George Washington University, Washington, D.C. [279]*
Chris A.G. van Eekelen, University of Nijmegen, The Netherlands [235]
Walther J. van Venrooij, University of Nijmegen, The Netherlands [235]*
Ethel Verney, George Washington University Medical Center, Washington, D.C. [111]
Bert Vogelstein, Johns Hopkins University School of Medicine, Baltimore, Maryland [169]*
Freidrich Wanka, University of Nijmegen, Nijmegen, The Netherlands [199]*
Valerie Zbrzezna, The Lindsley F. Kimball Research Institute of the New York Blood Center, New York, New York [223]

Preface

The nucleus, the largest organelle of the cell, has been investigated extensively, in particular as the locus of replication and transcription processes. Morphologically, few components or compartments can be recognized, much less isolated. The membranous subdivision, characteristic of many cytoplasmic organelles, is missing. Sonication could tear the nucleolus out of the general "chromatin" network. Enzymatic digestion and high-salt extraction result in residues with apparent structural properties. This latter fractionation has resulted in "artificially" derived components associated with a proliferating terminology and a variety of functions. In fact, the only nuclear function for which no evidence has as yet been found is transport.

Investigators from very different biological disciplines have tested the structures left over from the harsh nuclear extraction method used, with their respective technologies and against their hypotheses. Replication, transcription, hormone and carcinogen binding have been demonstrated by some and rejected as artifactual by others.

Surrounding the nucleus is a double membrane whose subcomponents are highly complex and seem to be continuous with, or part of, what has become known as the nuclear matrix. Investigations of this substructure are done with exchange mechanisms between the two major cellular compartments in mind.

At present, the research focused on an operationally defined structure seems to have reached the critical stage of acceptance, though not necessarily funding. The gathering at the Sugar Loaf Conference Center in Philadelphia was designed to bring investigators working on the structure together so that all could gain an overview of the immensely diverse ideas associated with this nuclear residue.

The steady increase of good publications may signal a growth of interest similar to that for the cytoskeleton. Formal similarities between those two structural components are beginning to spawn the first sweeping speculations.

The success of the meeting was due to the active and enthusiastic participation of all present, and hopefully a reassessment of the results by an expanded group will be possible after a few years.

Gerd G. Maul

ASPECTS OF A HYPOTHETICAL NUCLEOCYTOPLASMIC TRANSPORT MECHANISM

Gerd G. Maul

The Wistar Institute
36th Street at Spruce
Philadelphia, PA 19104

The two major compartments in the eukaryotic cell, the nucleus and the cytoplasm, are separated by a membranous barrier perforated by pores or annuli as already reported by the early electron microscopists (Callan, Tomlin 1950). This barrier separates the transcriptional and translational machinery. Evolutionarily, this separation must have been a complex sequence of events where many features of an exchange mechanism must have developed before the genetic material was enclosed. Dr. Cavalier-Smith addresses this problem (this volume), despite the limited information available about the different functions of diverse nuclear structures. Our background knowledge of nuclear activity increases rapidly, particularly in the field of DNA replication, transcription and processing of RNA. An impressive amount of detailed knowledge exists concerning translation and the processing of proteins, and for most of the processes, defined structures are known to be involved. However, the transport of the message from the nucleus and of proteins into the nucleus is associated with the nuclear pore complex only tentatively. This structure was originally considered as a gate, based on the assumptions that the cytosol was structureless and that macromolecular transport to and from the nucleus was by diffusion. The size of the pore complex precludes direct observation of particles passing through it, and only reconstructed morphological sequences of specialized nuclei, such as oocytes, spermatocytes or salivary gland cells (Stevens, Swift 1966) are available to suggest this passage. Nevertheless, evidence indicates an active transport mechanism, as proteins such as albumin cannot easily diffuse into the nucleus or do so only over much longer time periods than larger proteins specific for the nucleus. It is therefore surprising that this presumed transport mechanism has not received more attention.

The nuclear pore complex, which has been well described ultrastructurally (Franke et al. 1974; Maul 1977 for reviews), has always been the structural focus of this transport mechanism. Eight globular particles or traverse fibrils, central rings and radiating fibers were the dominant structures that suggested no other function than that of a passive sieve. However, if RNA synthesis and processing involves a solid phase (Herman et al. 1978; van Venrooij et al. 1981; Miller et al. 1978), and if no RNA leaves the nucleus when the envelope is experimentally torn (Feldherr 1980), then transport must also involve a solid phase system. This transport system must originate from deep within the nucleus and therefore structurally must be continuous or potentially continuous. Morphological examination of gross nuclear structure reveals channels that lead from single or small groups of pore complexes through the heterochromatin to areas of euchromatin. Lead precipitates indicate ATPase or GTPase activity along these channels and at the pore complex (Vorbrodt, Maul 1980). These enzymes are presumably essential for the energy necessary for transport (see Agutter et al. 1976; also Schumm, this volume). RNA can also be localized along these channels and euchromatin areas using preferential staining techniques (Franke, Falk 1970; Vorbrodt, Maul 1980) or DNA digestion (see "Nuclear Matrix," this volume). At a higher magnification, fine fibrils traverse the pore complex and can be followed in some cases deep into the nucleus (Fig. 1). Nuclear fractionation studies have revealed the so-called nuclear matrix, which if it exists, underlying the chromatin as a general support, would provide the solid phase framework for RNA synthesis and transport (see Pogo, this volume).

The traverse fibrils extend out into the cytoplasm, and may connect to cytoskeletal elements. Lenk et al. (1977) have reported on "free" ribosomes that become attached to a cytoplasmic network and that are active in protein synthesis only during their attached state. These traverse fibrils structurally connected to a cytoplasmic component would form a solid phase network throughout the cell. This, in turn, raises the possibility of continuity

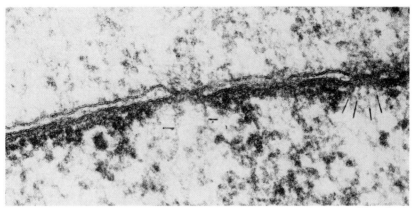

Figure 1. Cross-section of the nuclear envlope. The left pore complex shows the projection of traverse fibrils that are continuous from the cytoplasm deep into the nucleus. A polysome is attached on the cytoplasmic side. The right pore complex is sectioned so that one rim is included showing several of the traverse fibrils (arrows). Magnification 100,000X.

between the nuclear and cytoplasmic matrices, which has implications for transport and distribution of transcriptional products.

Such a speculative overview of active transport mechanisms has to be reduced to testable hypotheses. I present here alternate hypotheses, all centered on the pore complex. These hypotheses are based on observations and experimental data as well as analogous models of movement and transport. Some of these observations are depicted and briefly described in Figure 2. This figure should not be regarded as a model (which would suggest functions), but rather as a schematic representation of the various morphological alternatives.

Transport based on a mechanism other than diffusion would require a <u>vehicle</u> on or along which a product could be transported, an <u>attachment mechanism</u> to join the vehicle and the product, a <u>selection mechanism</u> for transport of specific macromolecules and a <u>release mechanism</u> at the end of transport. Some system of <u>energy generation</u> is required and, last but not least, a mechanism for <u>movement</u>.

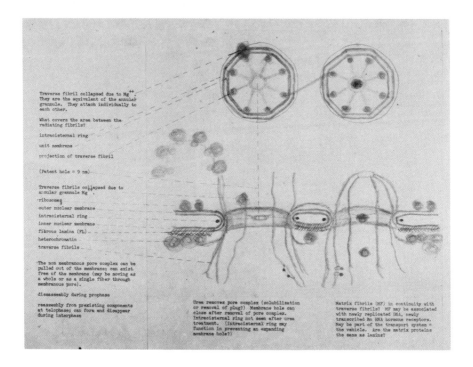

Figure 2. Schematic representation of the nuclear pore complex. The pore complex has an octagonal symmetry due to eight traverse fibrils. On the cytoplasmic side, they can collapse due to Mg^{++} ions and are then called annular granules. These fibrils can attach to each other and may aggregate isolated nuclear envelopes. Radiating fibrils may span the pore complex at its waist (left PC) or may be the projections of traverse fibrils (right PC). Central granules have been seen at different levels in the middle of the PC. An intracisternal ring surrounds the pore complex and is in need of a functional assignment. The non-membranous pore complex can easily be pulled out of the membranous pore and can exist free of the membrane in vivo. Polyribosomal rings on membranes have the same diameter as pore complexes if not longer than eight ribosomes.

If the pore functions as a sieve, this sieve must be constructed in the plane of the pore opening or across the direction of movement. The central ring and radiating fibers would provide the structure. Transport structures would be aligned with the direction of the movement of the macromolecules. Therefore, assuming that traverse fibrils (see Fig. 2) are the vehicle and the product is selected and attached, what then creates movement? Several models of movement can be proposed and tested experimentally.

1. The membrane at the waist of the pore complex could move a specifically attached traverse fibril by membrane flow. Nagel and Wunderlich (1977) (see also Schumm, this volume) reported a strongly reduced RNA transport in Tetrahymena at the membrane phase transition temperature of 18°C, but Clawson and Schmuckler (1978) report no break in the slope of RNP efflux rates over a large temperature range in isolated rat liver nuclei. In situ crosslinking of nuclear membranes using antibodies may be one approach to testing the membrane flow hypothesis.

2. Traverse fibrils are very adherent and attach to each other. They may attach to fibrous cytoplasmic components such as microfilaments, microtubules and microtrabeculae. Movements within the cytoplasm might cause the traverse fibrils to be pulled out of the nucleus. This passive movement could be interrupted by agents that disrupt the cytoskeleton or by those such as microinjected antibodies that "fix" it. movement generated in this way would be directional and thus not explain transport into the nucleus, unless, as suggested by Wunderlich and Herlan (1977), the nucleus changes in size and thus acts as a pumping mechanism.

3. Movement could be generated by assembly of the traverse fibril on one end and disassembly on the other end. However, the assembly site would require anchorage and rigidity (as do microtubules), and no evidence to date suggests this.

In these models of movements, the product is attached to the traverse fibril and moved "cable-car" fashion. However, if the traverse fibrils are not the vehicle, with its connotations of movement, but are instead as static as the electron micrographs would suggest, the product might move along a fixed track, "rail-car" fashion. The macromolecular analogue is that of ribosomes moving along a stationary messenger RNA strand where the ribosome represents the product to be moved and the messenger RNA, the traverse fibrillar track. [This image can also be applied to our first movement model. A structure may exist at the pore waist (analogue of a ribosome), through which the traverse fibril (analogous to mRNA) with its attached product would be moved.]

Transport is energy-dependent. In the rail-car hypothesis, the transported product has the function of the vehicle; it moves. In the cable-car hypothesis, the product is stationary with respect to the moving cable. The energy expenditure then is at different sites. This may be used to distinguish between the two hypotheses. It may be possible, for example, to crosslink the product to the cable or rail; if it is crosslinked to a cable, transport should proceed and ATP hydrolysis continue. If the cable can be crosslinked to another structure or other cables, transport should stop and ATP hydrolysis be reduced.

Specific antibodies against the NTPase involved in transport should not only block ATP hydrolysis, but also RNP efflux, confirming in a more specific way the results by Agutter (1976) who used other phosphatase inhibitors. Less specific inhibitors such as lectins also could be used on isolated nuclei to crosslink or block the carbohydrate groups of ATPase, resulting in RNP efflux inhibition. Even less specific would be the action of polycations such as cationized ferritin. They would crosslink the product to the vehicle as well as the vehicle to other structures. If movement is dependent on ATP hydrolysis and if the reverse is true, then ADP and P_i should not be released. This type of experiment could distinguish between the rail-car and cable-car hypotheses. Antibodies against the transport structures are the most powerful tool, as their specificity precludes the interruption of other metabolic processes. Initial attempts in this direction are described by Baglia and Maul (this volume).

When the basic problem of movement generation is better understood, one can then consider the possibilities of how the transport system works. For example, it is possible that traverse fibrils of nuclear pore complexes can move in and out of the nucleus independently, or that all eight of the traverse fibrils of one pore complex move in the same directon. One can think of the pore complex as a temporary assembly, designed for a quantum event only (they form and disappear very rapidly, see Maul et al. 1972; Coleman 1974). Alternatively, they could form and persist throughout the cell cycle or the life span of the cell. Individual pore complexes may have specialized functions, such as ribosomal transfer or protein influx, and some of the complexes may be equipped with special recognition mechanisms to select hormone receptor complexes in target cells and provide a direct link to the site of hormone action. Evolutionarily this type of selection of messages may have preceded total enclosure of the genetic material by a membrane. If a "track" or "cable" system exists within the nucleus, this system must represent a highly complex structural development that is dependent in part on the position of pore complex sites on the chromosomes (Maul 1977b) and reflected in a nonrandom distribution of the pore complex (see Maul et al. 1971). The transport mechanism itself, however, need not be formed during the reconstruction of the nucleus after mitosis.

Importantly, nuclear pore complex formation during interphase occurs from the nuclear side, involving the modification of heterochromatin to euchromatin (Maul 1971). Thus, the new transcription sites are close to the exit sites of the message, or they form a direct track to and through that short tunnel.

During mitosis, nucleoplasm and cytoplasm mix. At telophase, the nuclear envelope reforms around the highly condensed chromosomes excluding most of the cytoplasm. However, some of the soluble cytoplasmic proteins must remain in the nucleus without ill effect. The apparent selective accumulation of nuclear proteins could therefore easily be explained by selective binding from a soluble and diffusible pool. Like the argument for active transport, however, microinjected IgG and other soluble proteins do not penetrate the nucleus, but larger nuclear proteins do so very quickly. This argues for the existence of a selection mechanism.

If one assumes the existence of specific signals, one may have to select and test for a variety of existing models or systems, not all of which are mutually exclusive. For example, signals on proteins or RNP could exist for or against selection for transport. There could be a unique signal for transport between the compartments or product-specific signals. In the latter case, the selection mechanism could be unique for transport to specific sites in the nucleus either by a switching mechanism along the tracks if all pore complexes are equally receptive, or by recognition or acceptor type functions of individual pore complexes in direct continuity with such sites.

1) A counterpart to the codon-anticodon mechanism may exist for the recognition of RNP designed to be transported out of the nucleus or for the tRNA-amino acid recognition for protein transport in the opposite direction. Some transport could therefore occur during RNA processing with a release as a final step (see also Clawson and Schmuckler, this volume, for a detailed RNP transport model).

2) Transient covalent bonds like cysteine derivatives (thioesters), phosphate diesters (by which other functional groups are linked by the phosphate to serine and tyrosine) and other readily reversible modification reactions may make sites recognizable and ready for attachment. Wold (1981) lists O- and N-phosphates and N-acetates, the methyl esters of glutamate, the adenylyl-and uridylyl-O-tyrosine derivatives as well as poly-ADP-riboxyl derivatives as examples. If the pore complex has the ability to modify the proteins, one might be able to find the respective enzymatic function or the in vivo attachment site of the product for experimental analysis at the pore complex.

3) Other protein modifications include the addition of carbohydrate moieties. Lectins are apparently ubiquitous proteins essential for biological communication, and presumably they recognize and bind oligosaccharides. This recognition depends on oligosaccharide sequence, configuration, on glycosidic linkage, linkage position and branching as specific determinants (Goldstein, Hayes 1978). "If this is indeed the way the oligosaccharide structures are read in biological communication, it should be obvious, that an incredible varied and complex language can be derived from a relatively simple alphabet consisting only of the common monosaccharide constituents of glycoproteins and glycolipids." (Wold 1981). Though not all nuclear proteins contain a substantial amount of carbohydrate, small single groups would easily go undetected.

4) Proteins destined for the rough endoplasmic reticulum carry a signal peptide sequence and are recognized by a signal recognition protein (Walker, Blobel 1981), but very little is known about the recognition signals that result in transport to intracellular organelles, plasma membranes or extracellular space. The nucleus should be included in this separation process. A lamin has been reported to be synthesized as a larger protein than is actually found at the fibrous lamina, that is, after transport and functional integration (Blobel 1980). It is conceivable therefore that the proteins recognized for nuclear uptake are also genetically labeled by transcribed and translated signals, and are not recognized solely through post-transcriptional modification.

If neither the rail-car nor cable-car transport hypothesis is acceptable, the following construct seems possible. The pore complex could function primarily as a sieve with a specific gate that allows entry of large molecules if they carry the proper signal to induce a gate opening. The "transport" in such a case would be for a very short distance only. The enzymatic complement for the conformational changes and energy requirements may then be expected to be very specific to the pore complex. The search for such specific enzymatic properties of the pore complex or annulus may help to distinguish between what may be called the anal hypothesis and the rail- or cable-car hypothesis. This anal hypothesis is consistent with the ultrastructural observation that large RNP complexes are seemingly squeezed through the center of the annuli in specialized nuclei (Stevens, Swift 1966). Both types of hypotheses, the one based on a physical continuum, the other mostly on diffusion, can be combined in various ways. For example, solid-phase tracks or cables may be operative for transport in somatic nuclei, whereas a selective diffusion mechanism at the pore complex side could provide for the protein supply to the nucleus as well as for the hormonal and other signal transfers.

The pore complex is the ultrastructurally recognizable focus of our studies on the presumptive active and selective nucleocytoplasmic exchange of macromolecules. It should not be forgotten in this context that the pore complex often exists in large stacks or singly in the rough endoplasmic reticulum or within the nucleus where no transport function can be assumed. Should these accumulations of pore complexes not be an expression of excess building blocks resulting in self-assembly, they could function in as yet unimagined ways.

In conclusion, investigations concerned with nucleocytoplasmic exchange should center on experiments designed to discriminate between active or passive transport, the mode of movement and the selection process that regulates the traffic between the major compartments of the cell.

ACKNOWLEDGEMENT

Supported by grants GM 21615 and CA 10815 from the NIH.

Agutter PS, McArdle HJ, McCaldin B (1976). Nature 263:165.
Blobel G (1980). Personal communication.
Callan HG, Tomlin SG (1950). Proc R Soc London Ser B 137:367.
Cavalier-Smith (1982). These Proceedings.
Clawson GA, Smuckler EA (1978). Proc Natl Acad Sci USA 75:5400.
Coleman SE, Duggan J, Hackett RL (1974). Tissue and Cell 6:521.
Feldherr CM (1974). Exp Cell Res 85:271.
Feldherr CM (1982). These Proceedings.
Franke WW (1974). Int Rev Cytol 4:71.
Franke WW, Falk H (1970). Histochemie 24:266.
Goldstein JJ, Hayes CE (1974). Adv Carbohydr Chem Biochem 35:127.
Herman R, Wymouth L, Penman S (1978). J Cell Biol 78:663.
Lenk R, Ransom L, Kaufmann Y, Penman S (1977). Cell 10:67.
Maul GG, Price JW, Lieberman WM (1971). J Cell Biol 51:405.
Maul GG, Maul HM, Scogna JE, Lieberman MW, Stein GS, Hsu BY, Borun TW (1972). J Cell Biol 55:433.
Maul GG (1977b). J Cell Biol 74:492.
Maul GG (1977). Int Rev Cytol Supl 6:75.
Miller TE, Huang CY, Pogo AO (1978). J Cell Biol 76:675.
Nagel WC, Wunderlich F (1977). J Membrane Biol 32:151.
Paine PL, Moore LC, Horowitz SB (1975). Nature 254:109.
Schumm DE (1982). These Proceedings.
Stevens BY, Swift H (1966). J Cell Biol 31:55.

van Venrooij WJ, van Eekelen CAG, Mariman ECM, Reinders RY
  (1982). These Proceedings.
Walter P, Blobel G (1981). J. Cell Biol 91:557.
Wold F (1981). Ann Rev Biochem 50:783.
Wunderlich F, Herlan G (1977). J Cell Biol 73:271.

ATTEMPTED ISOLATION OF THE NUCLEAR PORE COMPLEX: EVIDENCE FOR A POLYDISPERSE COMPLEX OF POLYPEPTIDES

Robert P. Aaronson, Laura A. Coruzzi, and Kenneth Schmidt

Department of Microbiology
Mount Sinai School of Medicine of the City University of New York
New York, New York 10029

Nuclear pore complexes are clearly observed in nuclear envelopes isolated upon extraction of rat liver nuclei with high- or low-salt solutions. The pore complexes are absent, however, in nuclear membranes isolated in the presence of polyanions. Moderate concentrations of urea also appear to disrupt pore complexes. Polypeptide analyses fail to reveal any unique polypeptides, the presence or absence of which correlates with the presence or absence of pore complexes. These results support the hypothesis that the nuclear pore complex is a polydisperse complex.

INTRODUCTION

The nuclear envelope is a hallmark of eucaryotic cells (Stevens, Andre 1969; Franke 1970; Feldherr 1972; Wischnitzer 1973; Franke 1974; Wunderlich et al. 1976; Kay, Johnson 1973). It consists of two essentially concentric unit membranes, the outermost of which is, <u>in situ</u>, continuous with the rough endoplasmic reticulum and is studded with ribosomes on its outermost surface. The inner membrane is smooth and is in lateral continuity with the outer membrane at numerous circular interruptions, known as pores, in the concentric membranes. Situated within the pores are morphologically elaborate structures called nuclear pore complexes (Watson 1959; Franke 1970). Immediately beneath the inner membrane is a thin amorphous protein layer which completely surrounds the nucleus (Fawcett 1966; Kalifat et al. 1967; Patrizi, Poger, 1967) and with which the nuclear pore complexes may be closely associated (Aaronson, Blobel 1974; Aaronson, Blobel 1975; Dwyer, Blobel 1976).

The nuclear pores and pore complexes are the most distinctive morphological characteristics of the nuclear envelope. Several models (Stevens, Andre 1969; Abelson, Smith 1970; Feldherr 1972; Kessel 1973; Franke 1974) for the morphological structure of the pore complex have been proposed and are supported by excellent electron micrographs. The models for the pore complex, although irreconcilable in explicit detail, share several features: 1) uniform size and shape of the complex for any cell type at any particular time, 2) eight-fold radial symmetry, the axis of which is perpendicular to the plane of the envelope and coincident with the center of the pore, 3) a plane of symmetry at the level of continuity of inner and outer membrane, 4) extension beyond and overlapping the margins of the pore by the complex, and 5) a central granule representing material in transit through a central passageway of restricted size.

An underlying assumption, based on the presence of pore complexes in all eukaryotes examined to date, is that the pore complexes are responsible for some basic function(s) common to all cells. Two nontrivial explanations, then, for the discrepancies in observations emphasized by the various models would be that the differences arise owing to different physiological states of the various cell types or that pore complexes have evolved differently in different species. The pore complexes may now be sites of multiple functions and the apparatus required for such diversity would be necessarily complex and unlikely to be identical in the finer details. However, the basic assumption of a universal fundamental function for the nuclear pore complex seems reasonable. A consequence of this assumption is that some or all of the common features included in the models would be related to this fundamental function. Another consequence is that the composition and structure of all pore complexes should be related, reflecting this common fundamental function.

Considering the wealth of descriptive literature on the pore complex, there is little evidence concerning its chemical composition. Virtually all of the experiments characterizing the chemical nature of the pore complex are cytochemical, e.g., response to fixatives (Merriam 1961;

Mentre 1969), to hydrolytic enzymes (Merriam 1961; DuPraw 1965; Koshiba et al. 1970; Beaulaton 1968; Abelson, Smith 1970; Clerot 1968), to detergents (Aaronson, Blobel 1974; Aaronson, Blobel 1975; Dwyer, Blobel 1976), or to staining characteristics (Monneron, Bernhard 1969; Sheridan, Barnett 1969; Franke, Falk 1970). In summary, these data suggest, albeit weakly, that the pore complexes are largely composed of protein (or are kept intact by protein, or that heavy-metal staining requires the presence of protease-sensitive material, etc.). Krohne et al. (1978) have noted the presence of 2-4 polypeptides, in manually isolated nuclear envelopes from amphibian oocytes, insoluble in high- and low-salt buffers and Triton X-100, which they attribute to nuclear pore complexes. The pore complex-lamina cell fraction obtained from rat liver, which is highly enriched for structures resembling nuclear pore complexes, is approximately 95% protein and displays a relatively simple constellation of polypeptide species on sodium dodecyl sulfate (SDS) gels (Aaronson, Blobel 1975; Dwyer, Blobel 1976), but, none of these polypeptides has been identified as a pore complex constituent. Thus, in the course of our attempts at isolating the pore complex, we have performed polypeptide analyses of subfractions in order to correlate the presence or absence of specific polypeptides with the presence or absence of pore complexes. We report here that we have found that polyanions, urea, and guanidine HCl apparently disrupt pore complexes without releasing any detectable unique polypeptides. A positive interpretation of these results is that the pore complex is, in fact, a polydisperse aggregate of constituent polypeptides, each of which is present in (presumably) stoichiometric but virtually undetectable (by present techniques) quantities.

RESULTS AND DISCUSSION

Nuclei can be isolated from rat liver with a high degree of purity and retention of all of the structural features recognized in situ (Blobel, Potter 1966; Monneron et al. 1972), e.g., nucleoli, dense chromatin, inner and outer nuclear membranes, nuclear pore complexes, etc. Thus, rat liver nuclei are an excellent starting point for isolation of intact nuclear envelopes which would then serve as an intermediate fraction highly enriched for pore complexes. Indeed, at present, the only assay for nuclear pore complexes requires direct observation in the electron microscope. The nuclear envelope then, as a morphologically well-defined structure, is ideal for identification of pore complexes. In addition, the absence of pore complexes can be documented since empty nuclear pores can be identified.

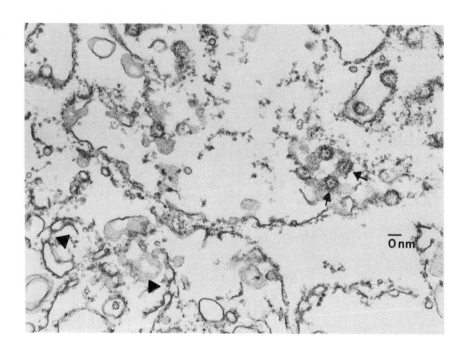

Fig. 1. A rat liver nuclear envelope fraction. The fraction was prepared by the very low salt method of Harris and Milne (1973). The small arrows point to nuclear pore complexes exhibiting central granules and substructure in the annular ring. The triangles indicate nuclear pore complexes in lateral view. The bar represents 100 nm. (Magnification, 40,000X.) The envelope fraction was prepared for electron microscopy as described previously (Aaronson, Blobel 1974). Briefly, a pellet of envelopes was fixed with glutaraldehyde, post-fixed with OsO_4, and stained en bloc with uranyl acetate before dehydration with ethanol and embedding in Epon. Thin sections were stained with both uranyl acetate and lead citrate.

Previous work has indicated that the majority of the internal components of the nucleus are associated via electrostatic interactions and that disruption of these associations with high concentrations of salt and/or by hydrolysis of the DNA using DNAse allows subsequent isolation of relatively intact nuclear envelopes (Harris 1978; Kay et al. 1972; Price et al. 1972; Monneron et al. 1972; Kashnig, Kasper 1969) in which nuclear pore complexes are well preserved. For instance, the envelope fraction (Fig. 1) contains a diffuse fine granular-fibrillar material but the bulk of the nuclear content, i.e., chromatin, nucleoli, etc., is absent. A thin, amorphous, electron-dense layer, the peripheral lamina, is closely apposed to the nucleoplasmic surface of the inner membrane. The parallel inner and outer membranes of the envelopes are continuous at the nuclear pore margins and pore complexes are evident in en face views as 85-100 nm diameter circular objects, virtually all of which contain a distinct central granule and exhibit annular substructure. Lateral views of pore complexes are more difficult to identify unambiguously since positive identification requires that both the inner and outer nuclear membranes be intact and the complex be perfectly oriented in the section (rim of pore problem). Presumptive lateral views of pore complexes are, however, observable (Fig. 1). Similarly, nuclear envelopes isolated using high-salt procedures also exhibit nuclear pore complexes (Kay et al. 1972; Aaronson, Blobel 1975) and are relatively free of internal nuclear components (not shown).

Biochemical Models of the Nuclear Pore Complex

Although numerous morphological models for the pore complex have been published (see above), there has been little speculation about their biochemical organization. In view of the large size of the pore complex and the apparent importance of protein to the integrity of the pore complex, two extreme models for the biochemical composition of the complexes can be envisioned and are presented in Figure 2. In the simple model, which is analogous to the structure of microtubules, the pore complexes are composed predominantly of a single or a very few closely related polypeptides which polymerize to form the basic structure. Associated proteins (by analogy with the microtubule-associated proteins) may play an important physiological role but are not required for the formation of the basic structure. In contrast, a

Fig. 2. Models of the nuclear pore complex.

Simple (microtubule-like)	Compound (ribosome-like)
The pore complex is an aggregate of one (or very few) species of polypeptide (or RNA), of which there are identical copies in each pore complex.	The pore complex is an aggregate of a large number of distinct molecular species and there is one copy of each per pore complex or subunit.

compound or complex model, relying on analogy with the ribosome, assumes that the pore complex is comprised of a polydisperse stoichiometric set of polypeptides (and perhaps RNA, etc.). Again, associated proteins may play an important physiological role but need not be present in stoichiometric amounts and may be dissociated under mild conditions.

One consequence of the simple model is that a significant amount of a single (or a very few) polypeptide(s) should be detected upon analysis of the polypeptides of a fraction highly enriched for nuclear pore complexes. An estimate of the amount of such a polypeptide in, for instance, a nuclear envelope fraction from rat liver can be obtained if the following assumptions are made:

There are at least 8 annular subunits with diameters of approximately 20 nm (Franke 1970).

The density of an "average" protein is 1.35 gm/cm^3.

There are approximately 3,500 nuclear pores per rat hepatocyte nucleus (Maul 1977).

One obtains approximately 450×10^6 nuclei and 1.5 mg of nuclear envelope protein from a 7 g liver (Aaronson 1977).

The yield of nuclear envelope material is approximately 50% of that present in intact nuclei (Wunderlich et al. 1976).

The first two assumptions predict the minimum mass of a pore complex to be approximately 0.5×10^{-16} g. This appears to be a very conservative estimate compared with the value of 5.2×10^{-16} reported from electron scattering measurements (Dupraw, Bahr 1969). Thus, one expects, at a minimum, approximately:

$(450 \times 10^6) \times (3,500) \times (0.5 \times 10^{-16}) \times 50\% = 40$ µg of pore complex protein/rat liver, or approximately 2.5% of the total nuclear envelope protein may be pore complex protein.

We can easily analyze, with good resolution, 50 µg of protein from a rat liver nuclear envelope fraction in one-dimensional denaturing polyacrylamide gels where the minimum amount of a homogeneous polypeptide detectable is below 0.5 µg. Thus, if 1.3 µg (i.e., 2.5% of 50 µg) of protein were present as a homogeneous polypeptide, as assumed in the simple model, a distinct detectable band should be observed (unless it is obscured by other bands). On the other hand, if the 1.3 µg were stoichiometrically distributed as 75-100 distinct polypeptides, as assumed by the complex model, each would be well below the detection limits of these gels.

Polypeptide Composition of Nuclear Envelope Fractions

The polypeptide profiles of nuclear envelope fractions obtained in the presence of low- or high-salt are relatively simple and virtually identical except for the occasional presence of histones (see Fig. 3). The majority of the polypeptides are present in clusters in the range 65,000-70,000 and 48,000-51,000. The latter cluster is solubilized by nonionic detergents (Aaronson, Blobel 1975; Scheer et al. 1976) and thus cannot be major constituents of the pore complex (Aaronson, Blobel 1974, Aaronson, Blobel 1975; Scheer et al. 1976; Krohne et al. 1978).

The polypeptides which comprise the cluster in the molecular weight region 65,000-70,000 have been immunologically localized to the nuclear peripheral lamina (Gerace et al. 1978; Krohne 1981). There are relatively few other polypeptides which are repeatedly present in sufficient amounts to be considered likely constituents of the pore complexes. Thus, it seems likely that the pore complex may be polydisperse with respect to polypeptide composition.

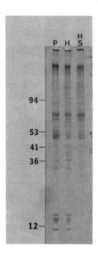

Fig. 3. Polypeptide analysis of rat liver nuclear envelope fractions prepared by three different methods. The fractions were solubilized with 0.2% SDS, reduced and alkylated, and subjected to electrophoresis in a linear 10-15% gradient polyacrylamide gel in the presence of SDS using a discontinuous buffer system (Maizel 1969; Aaronson, Blobel 1974). P, the nuclear envelope fraction isolated in the presence of polyvinylsulfate (Aaronson 1977); H, envelopes isolated upon incubation of nuclei with DNase I in very low salt conditions (Harris, Milne 1973); HS, envelopes isolated upon extraction of DNase I-treated nuclei with 1 M NaCl (Aaronson, Blobel 1975; Dwyer, Blobel 1976; Aaronson 1977). The numbers at the left indicate the molecular weight ($\times 10^{-3}$) of protein standards electrophoresed in an adjacent slot.

Apparent Dissolution of Nuclear Pore Complexes by Polyanions

Agents other than DNAse I and high-salt affect electrostatic interactions in the nucleus in vitro. It has been shown that relatively low concentrations of polyanions cause striking changes in nuclear ultrastructure as well as in the accessibility of the DNA in chromatin (Ansevin et al. 1975; Arnold et al. 1972). Heparin, a sulfated glycosaminoglycan, has been utilized to isolate nuclear envelopes (Bornens 1973; Bornens, Courvalin 1978; Hildebrand, Olinaka 1976). Electron micrographic examination of an envelope fraction obtained in the presence of polyvinylsulfate (Aaronson 1977) shows a relatively homogeneous pellet of empty double membrane nuclear envelopes, fragments of envelopes, and single membrane vesicles (not shown). The smaller fragments were predominant near the top of the pellet whereas the more intact nuclear ghosts predominated at the bottom. Occasional clusters of small single-membrane vesicles were observed together with fragments of double membrane envelope throughout the pellet. Many, although not all, of the single-membrane vesicles exhibited ribosomes on their surface and had diameters up to 1 μm, indicating that they represented large pieces of the outer nuclear membrane.

The larger double membrane nuclear envelopes exhibited contour lengths greater than 20 μm, implying that they were essentially the entire nuclear envelopes of individual nuclei. Complete circular continuity of even these large fragments was very rarely observed, indicating that none of the nuclear envelopes remained completely intact during this treatment. At higher magnifications (Fig. 4 and 4 insert), the double membrane fragments can be examined more closely. It can be seen that ribosomes are attached to one side of one of the membranes allowing clear identification of outer and inner nuclear membranes. Numerous nuclear pores can be identified in en face and lateral views. Most of the pores, in the en face views, appear to be empty although a few appear to contain complete nuclear pore complexes or at least central granules and peripheral granules. Lateral views (insert) indicate that the nuclear pores are largely empty although some amorphous material is occasionally present at the margins. Except for the presence of ribosomes on the outer nuclear membrane, both membranes appear identical in this preparation. Finally, it is worth noting that, as expected (Maul 1977), incubation of nuclei with DNAse I alone under the present conditions leaves the nuclear pore complexes intact.

In as much as nuclear pore complexes were not present in nuclear envelopes exposed to polyvinylsulfate, the polypeptide composition of nuclear envelopes isolated in its absence and in its presence was determined (Fig. 3). Whether the polypeptides were visualized by Coomassie blue staining or by a much more sensitive Cu^{+2}/Ag^+ method (Merril et al. 1979), there were no polypeptides present in the envelopes containing pore complexes that were absent from the envelopes in which pore complexes could not be observed. There are some extra polypeptides in the polyvinylsulfate-derived envelopes; however, the yield of nuclear material in this fraction was higher than for the envelope fraction obtained in the absence of polyvinylsulfate so that any extra polypeptides observed may simply reflect the presence of contaminants. Similar results were obtained if polyphosphate (degree of polymerization = 200) was used in the preparation of envelopes (not shown).

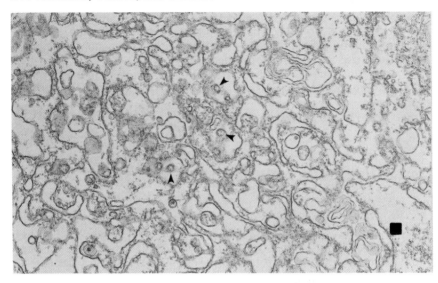

Fig. 5. Nuclear envelopes extracted with 2 M urea. Arrows indicate empty nuclear pores viewed en face. The bar represents 0.5 µm. Magnification 30,000X.

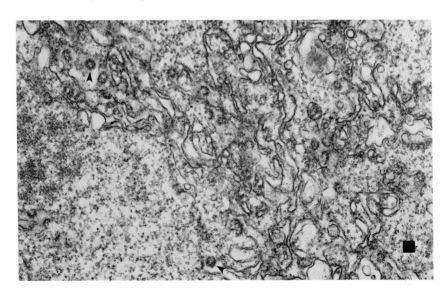

Fig. 6

The larger double membrane nuclear envelopes exhibited contour lengths greater than 20 µm, implying that they were essentially the entire nuclear envelopes of individual nuclei. Complete circular continuity of even these large fragments was very rarely observed, indicating that none of the nuclear envelopes remained completely intact during this treatment. At higher magnifications (Fig. 4 and 4 insert), the double membrane fragments can be examined more closely. It can be seen that ribosomes are attached to one side of one of the membranes allowing clear identification of outer and inner nuclear membranes. Numerous nuclear pores can be identified in <u>en</u> <u>face</u> and lateral views. Most of the pores, in the <u>en</u> <u>face</u> views, appear to be empty although a few appear to contain complete nuclear pore complexes or at least central granules and peripheral granules. Lateral views (insert) indicate that the nuclear pores are largely empty although some amorphous material is occasionally present at the margins. Except for the presence of ribosomes on the outer nuclear membrane, both membranes appear identical in this preparation. Finally, it is worth noting that, as expected (Maul 1977), incubation of nuclei with DNAse I alone under the present conditions leaves the nuclear pore complexes intact.

In as much as nuclear pore complexes were not present in nuclear envelopes exposed to polyvinylsulfate, the polypeptide composition of nuclear envelopes isolated in its absence and in its presence was determined (Fig. 3). Whether the polypeptides were visualized by Coomassie blue staining or by a much more sensitive Cu^{+2}/Ag^+ method (Merril et al. 1979), there were no polypeptides present in the envelopes containing pore complexes that were absent from the envelopes in which pore complexes could not be observed. There are some extra polypeptides in the polyvinylsulfate-derived envelopes; however, the yield of nuclear material in this fraction was higher than for the envelope fraction obtained in the absence of polyvinylsulfate so that any extra polypeptides observed may simply reflect the presence of contaminants. Similar results were obtained if polyphosphate (degree of polymerization = 200) was used in the preparation of envelopes (not shown).

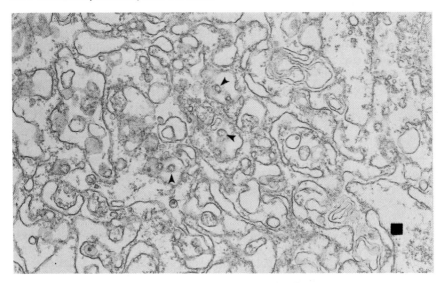

Fig. 5. Nuclear envelopes extracted with 2 M urea. Arrows indicate empty nuclear pores viewed en face. The bar represents 0.5 μm. Magnification 30,000X.

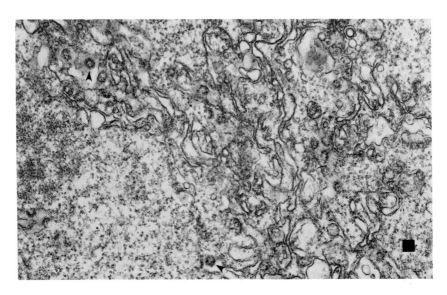

Fig. 6

Fig. 6 (Previous page). Nuclear envelopes extracted with 0.8 M urea. Arrowheads indicate nuclear pore complexes viewed en face. Central granules are not seen. The annular material does not exhibit the same substructure as in untreated envelopes (Fig. 1). The bar represents 0.5 μm. Magnification 30,000X.

Unfortunately, the nuclear peripheral lamina and outer membrane-bound ribosomes may be, at least partially, removed or disrupted under some conditions. The annular material, although present at a urea concentration of 2 M, seems to be structurally altered. The nuclear membranes appear to be relatively unaffected by these treatments.

Polypeptide Analysis of Nuclear Envelope Treated with Urea

The differential sensitivity of the nuclear pore complex components to urea thus permitted controlled disassembly of the nuclear pore complex. Therefore, we attempted to correlate the polypeptide profiles of urea-treated envelopes with the morphology of the pore complex during disassembly. Figure 7 is an SDS-polyacrylamide gel of nuclear envelope fractions obtained upon treatment of envelopes with urea. The gel was stained with a highly sensitive Cu-Ag stain (Merril et al. 1979) which pointed up a large number of polypeptides including histones which were previously undetected in nuclear envelope fractions (compare Fig. 3).

As the concentration of urea increased there was an apparent partial and concerted solubilization of all the envelope polypeptides. Moreover, there were none which were disproportionately solubilized as would have been expected from the apparent disruption of pore complexes at 0.8 and 2 M urea.

Similar morphological and biochemical results were obtained with the chaotropic salt guanidine HCl (not shown). The critical concentration range for dissection and dissassembly of the pore complex was between 0.3 and 0.8 M guanidine. Thus, using two denaturing compounds, urea and guanidine HCl at moderate concentrations, we were able to disrupt pore complexes with some selectivity. In neither

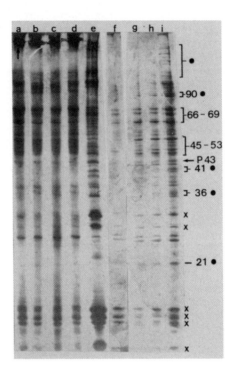

Fig. 7. Analysis of the effect of urea on the polypeptide composition of rat liver nuclear envelopes. Lane a represents the control envelope fraction. Lanes b, c, and d are the envelope fractions treated with 0.5 M, 0.8 M, and 2.0 M urea, respectively. Lane e represents the whole rat liver nuclear fraction. Lane f is the control rat liver nuclear envelope supernatant. Lanes g, h, and i represent the supernatants of the envelope fractions treated with 0.5 M, 0.8 M and 2.0 M urea, respectively. The supernatant lanes f through i were applied at equal relative volumes. Numbers to the right of lane i indicate the estimated interpolated molecular weight ($\times 10^3$) of the indicated polypeptide based on its mobility relative to standard proteins electrophoresed in a parallel slot. Electrophoresis was performed as in Figure 3.

case, however, were we able to correlate the presence or absence of a single or even a limited number of polypeptides with the presence or absence of identifiable pore complexes.

SUMMARY

Two extreme models for the polypeptide composition of the nuclear pore complex have been proposed. One assumes a relatively simple composition in which one or a very few polypeptides polymerize to form the pore complex. Calculations based on this assumption predict that such a polypeptide or small constellation of polypeptides should be readily detectable in nuclear envelope fractions. Unfortunately, all of the reproducible prominent polypeptides in the envelope fraction have been assigned to structures other than the pore complex. Polypeptide analyses of nuclear fractions exhibiting pore complexes and fractions in which pore complexes are apparently absent or partially disrupted failed to reveal candidate polypeptides the presence or absence of which correlated with the presence or absence of pore complexes. This evidence suggests to us that minor polypeptides or even presently undetected polypeptides are constituents of the pore complex and that its structure is, in fact, very complex.

ACKNOWLEDGMENTS

We would like to acknowledge the expert technical assistance of Ms. Denise McHale and Mr. Edwin Woo. This work was supported by grants from the National Institutes of Health (GM 21950 and GM 00278) and the National Science Foundation (PCM77-26587).

Aaronson RP (1977). Isolation of nuclear proteins associated with the nuclear pore complex and the nuclear peripheral lamina of rat liver. Meth Cell Biol 13:337.
Aaronson RP, Blobel G (1974). On the attachment of the nuclear pore complex. J Cell Biol 62:746.
Aaronson RP, Blobel G (1975). Isolation of nuclear pore complexes in association with a lamina. Proc Natl Acad Sci USA 72:1007.
Abelson HT, Smith GH (1970). Nuclear pores: The pore-annulus relationship in thin section. J Ultrastr Res 30:558.

Harris JR (1978). The biochemistry and ultrastructure of the nuclear envelope. Biochim Biophys Acta 515:55.

Kalifat SR, Bouteille M, Delarue J (1967). Etude ultrastructurale de la lamelle dense observee au contact de la membrane nucleaire interne. J Microscopie 6:1019.

Kashnig DM, Kasper CB (1969). Isolation, morphology and composition of the nuclear membrane from rat liver. J Biol Chem 244:3786.

Kessel RG (1973). Structure and function of the nuclear envelope and related cytomembranes. Prog Surf Membr Sci 6:243.

Koshiba K, Smetana K, Busch H (1970). On the ultrastructural cytochemistry of nuclear pores in Novikoff hepatoma cells. Exp Cell Res 60:199.

Krohne G (1982). This volume.

Krohne G, Franke WW, Scheer U (1978). The major polypeptides of the nuclear pore complex. Exp Cell Res 116:85.

Maizel JV (1969). Acrylamide gel electrophoresis of proteins and nucleic acids. In Habel K, Salzman NP (eds): "Fundamental Techniques in Virology", New York: Academic Press, p 334.

Maul GG (1977). The nucleus and the cytoplasmic pore complex: Structure, dynamics, distribution, and evolution. In Bourne GH, Danielli JF, Jeon WW (eds): Int Rev Cytol Suppl 6, New York: Academic Press, p 75.

Mentre P (1969). Presence d'acide ribonucleique dans l'anneau osmiophile et le granule central des pores nucleaires. J Microscopie 8:51.

Merril CR, Switzer RC, van Keuren ML (1979). Trace polypeptides in cellular extracts and human body fluids detected by two-dimensional electrophoresis and a highly sensitive silver stain. Proc Natl Acad Sci USA 76:4335.

Monneron A, Bernhard W (1969). Fine structural organization of the interphase nucleus in some mammalian cells. J Ultrastr Res 27:266.

Patrizi G, Poger M (1967). The ultrastructure of the nuclear periphery: The zonula nucleum limitations. J Ultrastr Res 17:127.

Price MR, Harris JR, Baldwin RW (1972). A method for the isolation and purification of normal rat liver and hepatoma nuclear "ghosts" by zonal centrifugation. J Ultrastr Res 40:178.

Scheer U, Kartenbeck JE, Trendelenburg MF, Stadler J, Franke WW (1976). Experimental disintegration of the nuclear envelope. J Cell Biol 69:1.

Stevens BJ, Andre J (1969). The nuclear envelope. In Lima-de-Faria A (ed): "Handbook of Molecular Cytology", p 837.
Watson ML (1959). Further observations on the nuclear envelope of the animal cell. J Biophys Biochem Cytol 6:147.
Wischnitzer S (1973). The ultrastructure of the nucleus and nucleocytoplasmic relations. Int Rev Cytol 10:137.
Wunderlich F, Berezney R, Kleinig H (1976). The nuclear envelope: An interdisciplinary analysis of its morphology, composition, and functions. In Chapman D, Wallach DFH (eds): "Biological Membranes", New York: Academic Press, p 241.

PROTEIN KINASE AND PHOSPHATASE REACTIONS OF THE NUCLEAR ENVELOPE

Khalil Ahmed and Randolph C. Steer

Toxicology Research Laboratory,
Department of Laboratory Medicine and
Pathology, University of Minnesota,
Veterans Administration Medical Center
Minneapolis, MN 55417

The nuclear envelope (NE) is of particular interest for a variety of reasons. Its structure is generally analogous to that of other biological membranes, with the additional feature that it contains characteristic "pore complexes", proteinaceous entities which have been implicated in nucleo-cytoplasmic exchange. Since the NE is a barrier separating the nuclear and cytoplasmic compartments, it is logical to examine the possible role of this membrane system and its associated pore complexes in regulating the movement of substances between these compartments. In addition, the envelope must undergo dynamic changes coincident with growth and cell division. A variety of enzyme activities has been ascribed to the envelope and considerable effort has been directed to studies on the structure and chemical composition of this organelle (for reviews see Franke, Scheer 1974; Kasper 1974; Harris 1978).

The cell nucleus is enriched in phosphoproteins and their associated enzymes; distinct protein kinase reactions appear to be localized in specific subnuclear fractions, and have been implicated in the control of transcription (for references see e.g. Ahmed 1975; Ahmed, Wilson 1978; Ahmed et al. 1979; Ahmed et al. 1981). However, until recently, no report had been published on the existence of such enzyme-catalyzed reactions in the nuclear envelope (Steer et al. 1979a; Lam, Kasper 1979; Steer et al. 1980; Agutter et al. 1979a; McDonald, Agutter 1980; Clawson et al. 1980a,b; Kletzien 1981; Veneziale et al. 1981). Phosphorylation of various other membranes has been demonstrated and may be centrally involved in a number of physiological processes (see Ahmed 1975; Greengard 1978; Rubin, Rosen 1975).

A nucleoside triphosphatase present in NE appears to be related to movement of RNA from nucleus to cytoplasm (Agutter et al. 1979b). It shall be interesting to determine whether this ATPase might reflect the combined actions of protein kinase and phosphatase reactions analogous to a number of other examples (see Judah, Ahmed 1964; Ahmed 1975). In this communication, we describe recent work from this laboratory demonstrating the presence and general catalytic properties of protein kinase and phosphatase reactions associated with the mammalian NE.

PREPARATION OF NUCLEAR ENVELOPE AND PORE COMPLEX-LAMINA

A prerequisite to the study of envelope-associated protein kinase and phosphatase reactions is availability of an envelope preparation minimally contaminated by ribosomes or chromatin, since these fractions of the cell also contain enzymes and substrates for protein phosphorylation and dephosphorylation (see e.g. Ahmed 1975; Ahmed, Wilson 1978; Ahmed et al. 1979, 1981; Rubin, Rosen 1975). The following procedure, which is a modification of that described by Kay et al. (1972) and incorporates additional improvements in the previously described procedure (Steer et al. 1979a,b), has been adopted for routine use in our laboratory. It provides highly reproducible preparations of purified NE from rat liver.

Washed intact nuclei from 20 g of rat liver pulp were suspended in a medium consisting of 0.1 mM $MgCl_2$, 5 mM 2-mercaptoethanol, 1 mM phenylmethylsulfonyl fluoride, 150 µg/ml of DNase I, and 10 mM Tris-HCl, pH 8.6 at 37°C. The suspension (5 ml final volume) was allowed to stand at 23°C for 20 min at the end of which 20 ml of ice-cold H_2O was added, and the material was centrifuged for 20 min at 29,000 x g at 4°C. The pellet was resuspended in the above medium except that 50 µg/ml DNase I was added and the pH was 7.4. This digestion was carried out at 23°C for 30 min and was terminated by 8 ml of ice-cold H_2O, mixed thoroughly, and centrifuged as indicated above. The pellet was then suspended in 16 ml of 0.25 M sucrose, and EDTA (pH 7.45) was added to a final concentration of 20 mM prior to layering on a discontinuous sucrose density gradient for centrifugation at 80,000 x g for 16 hr at 4°C.

Approximately 90% of the recovered NE was harvested from the 1.28 g/ml - 1.19 g/ml sucrose interface while the remainder was collected at the boundary between 0.25 M and 1.19 g/ml sucrose. The protein composition (as determined by gel electrophoresis) and enzymic properties of the two envelope fractions were almost identical. Transmission electron microscopy of the envelope samples revealed well-preserved symmetrically apposed membrane sheets with evidence of pore complexes. Subfractionation of the purified intact envelope to isolate a pore complex-lamina preparation was carried out according to Dwyer and Blobel (1976). These techniques provided rat liver NE and pore complex-lamina preparations which were minimally contaminated by DNA and RNA, suggesting that chromatin and ribosomal components were effectively removed during the isolation procedures (Table 1). For comparison, some data from studies on guinea pig seminal vesicle epithelium NE are included to emphasize that the composition of envelope preparations may vary depending on the tissue source, and that well-defined modifications of existing isolation methods might be necessary in order to isolate desirable preparations from different tissues (Veneziale et al.

Table 1

Composition of Nuclear Envelope and Pore Complex-Lamina

Preparation	µg/gm of liver pulp[a]			µg/gm of epithelium[b]		
	Protein	DNA	RNA	Protein	DNA	RNA
Nuclei	2080	874	149	4380	1730	315
Nuclear envelopes	108	6	11	181	1.4	5.0
Pore complex-lamina	18	0.07	0.19	-	-	-

[a] Steer et al., unpublished observations.
[b] Data from Veneziale et al. (1981).

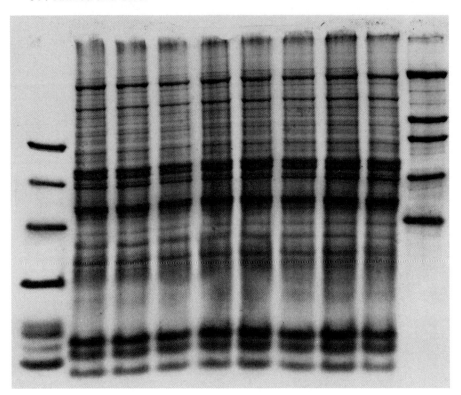

Fig. 1. Gel electrophoretic profile of rat liver nuclear envelope proteins from eight separate NE preparations. Electrophoresis was carried out according to Veneziale et al. (1981) except that 50-75 μg of envelope protein was applied to each well. Molecular weights of protein standards (far left) were from top to bottom: 94K, 68K, 43K, 30K, 21K and 14.3K daltons; and (far right): 200K, 130K, 94K, 68K and 43K daltons.

1981). That highly reproducible envelope preparations may be obtained by the above procedures is evidenced by the similarity of electrophoretic profiles of eight separate liver envelope preparations (Fig. 1). It should be noted

Table 2

Properties and Requirements of NE-Associated Protein Kinase Reactions[a]

		Protein substrates	
	NEP	DPV	LRH
Km for protein substrate	–	0.30	0.36
Km for ATP (μM)	–	14	20
pH Optimum	6.5-8.5	7.5	7.8
NaCl (mM)	0	200	0
$MgCl_2$ (mM)	10-12	10	6-16
	Activity, % of Complete Control System		
+ protein kinase inhibitor (100 μg/ml) (400 μg/ml)	115 134	96 95	101 98
+ cAMP (1.0 μM) (10.0 μM)	101 100	95 98	115 145
+ cGMP (1.0-10 μM)	98	98	108
+ Spermine (1-5 mM)	100	–	–
+ Triton X-100 (2%, v/v)	22	31	29
+ DOC (0.1%, w/v)	73	81	68

[a] Steer et al., unpublished observations.

that the envelope contains multiple proteins in addition to the conspicuous 'triplet band' of proteins in the 68K - 78K molecular weight range (Aaronson, Blobel 1975).

PROPERTIES OF NUCLEAR ENVELOPE PROTEIN KINASE REACTIONS

The general properties of rat liver NE-associated protein kinase reactions are shown in Table 2. Similarities exist among the properties towards the three substrates tested, although some major differences are apparent. In particular, enzyme activity towards lysine-rich histone (LRH) and endogenous nuclear envelope proteins (NEP) did not require monovalent salt, whereas that towards partially dephosphorylated phosvitin (DPV) was maximal in the presence of 200 mM NaCl. Divalent cations other than Mg^{2+} were generally inhibitory, however, a stimulative effect of $CoCl_2$ was observed (Steer et al. 1980). Kinase activity towards DPV and LRH was not inhibited by the cyclic AMP-dependent protein kinase inhibitor from rabbit skeletal muscle; however, some stimulation of the protein kinase activity towards LRH was observed in the presence of cyclic AMP. This suggests that the protein kinase active towards this substrate was present as a holoenzyme. Since the effect of cyclic AMP was not very dramatic, it would also appear that some of this activity was due to a cyclic AMP-independent protein kinase. The inhibitor protein stimulated the phosphorylation of NEP. Detergents such as Triton X-100 and sodium deoxycholate (DOC) inhibited kinase activity when added directly to the assay; this inhibition was reversed on removal of the detergents by repeated washings. This suggests that protein kinase activities in NE are not present in a 'latent' form. Agents such as polyamines (spermine or spermidine at 1-5 mM) and calmodulin (at various concentrations in the presence or absence of Ca^{2+}) did not influence enzyme activity under the conditions employed.

A study of the time course of protein phosphorylation by the envelope-associated kinase activity indicated that activity towards DPV proceeded at a linear rate for 60 min, that towards LRH remained linear through 30 min, but phosphorylation of NEP did not remain linear beyond 15 min of reaction time. The extent of phosphorylation achieved in various proteins of the NE is shown in Figure 2, which represents an autoradiogram of $[^{32}P]$-labeled proteins from separate envelope preparations shown in Figure 1. In accord with the observation of Lam and Kasper (1979), enrichment of phosphorylation in the region of the triplet band was noted. However, labeling of other proteins also occurred. Interestingly, the region where histones and lower molecular weight non-histones should be present, contained essentially no radioactivity.

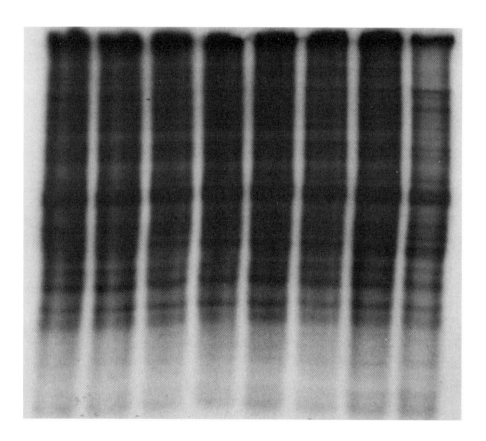

Fig. 2. Autoradiogram of gel electrophoretic profile of [^{32}P]-labeled NEP from eight separate envelope preparations. NEP was phosphorylated using [γ-^{32}P]ATP under optimal reaction conditions, electrophoresed (Fig. 1), and exposed to Kodak "No-Screen" x-ray film for 24 hr. The gels shown are duplicates of those in Figure 1.

Fractionation of intact NE to yield a preparation of pore complex-lamina, and its subsequent analysis as above have shown that a substantial part of the total envelope protein kinase activity is associated with the pore-lamina. As shown in Table 3, significant protein kinase activity was measured towards endogenous proteins of the pore complex-lamina, whereas only a scant activity towards LRH, and none towards DPV, was recovered. Thus it appears that the various protein kinase reactions associated with the NE may be due to multiple enzymes, and that some of them are either not localized in the pore complex-lamina fraction or are removed during its isolation.

Table 3

Protein Kinase and Phosphatase Associated with Nuclear Envelope and Pore Complex-Lamina

Preparation	Relative enzyme activities[a]				
	Kinase toward substrates			Phosphatase toward substrates	
	NEP	DPV	LRH	^{32}P[NEP]	[^{32}P]LRH
Nuclear envelope	1.5	26	2.2	0.53	8.3
Pore complex-lamina	2.3	0	0.4	0	8.4

[a] nmoles ^{32}P/mg of protein/hr.

PROPERTIES OF NUCLEAR ENVELOPE PROTEIN PHOSPHATASE REACTIONS

The observation that incorporation of ^{32}P from [γ^{32}P]ATP into NEP was increased by NaF (Fig. 3, panel A) prompted us to examine whether this event resulted from inhibition of an endogenous protein phosphatase reaction. Further evidence for the phosphatase was provided when it was shown that addition of EDTA to block the kinase reaction resulted in a steady loss of ^{32}P from previous labeled NEP (Fig. 3, panel B). The apparent rate of the protein phos-

Table 4

Properties and Requirements of the Protein Phosphatase Reactions[a]

pH Optimum	Protein Substrates	
	$[^{32}P]$PV	$[^{32}P]$LRH
	Activity, % of control with substrate alone	
+ NaCl (100 mM)	84	136
(200 mM)	71	77
+ 2-SH-EtOH (1 mM)	67	136
+ EDTA (1 mM)	49	163
+ NaF (16 mM)	63	45
+ DOC (0.1%, w/v)	-	-
+ NH_4-molybdate (1 mM)	-	-
+ Poly-A (100 μg/ml)	-	-
+ Poly-G (100 μg/ml)	-	133
+ $MgCl_2$ (1 mM)	103	115
+ $MnCl_2$ (1 mM)	123	110
+ $CaCl_2$ (1 mM)	110	108
+ $ZnCl_2$ (1 mM)	110	10
+ $CuCl_2$ (1 mM)	62	20

[a] Steer et al. (1979b) and Steer et al., unpublished observations.

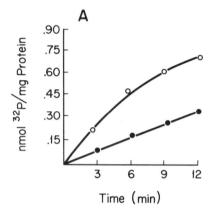 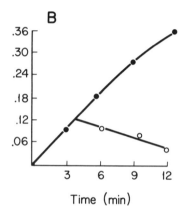

Fig. 3. Demonstration of an endogenous protein phosphatase activity associated with NE. A) Effect of incubation of NE in the presence (O) or absence (●) of 20 mM NaF on the rate of ^{32}P incorporation from $[\gamma-^{32}P]ATP$ into NEP. B) Effect of addition of 15 mM EDTA to the reaction at 4 min after the initiation of phosphokinase reaction. The discharge of radioactivity (O) represents phosphatase reaction. The rates of reactions shown are somewhat different from those reported previously (Steer et al. 1979b); this is because a more highly purified NE preparation, as described above, became available for these experiments.

phatase reaction under these conditions appears less than that of the kinase. Envelope-associated protein phosphatase activity was also demonstrated using [^{32}P]-labeled exogenous phosphoprotein substrates. Some general properties of the phosphatase reactions toward [^{32}P]PV and [^{32}P]LRH are shown in Table 4, which indicates that the separate reactions had several different properties. In general, divalent cations were without effect, however, $CuCl_2$ and $ZnCl_2$ inhibited the activity towards LRH. 1.0 mM EDTA enhanced phosphatase activity towards LRH, but inhibited that towards phosvitin (PV). Thus, it appears that the former activity, in contrast to the latter, may not require divalent cations. Several other differences were also observed, e.g., NaCl at 100 mM stimulated LRH phosphatase but was somewhat inhibitory towards PV phosphatase. At 200 mM, both activities were reduced. Also, NaF inhibited the activity toward LRH more potently than that toward PV. These results are evidence for multiple protein phosphatases associated with the NE. Agutter et al. (1977) have shown that ATPase associated with liver envelope is stimulated by polyadenylic acid (poly-A) and polyguanylic acid (poly-G). It is, therefore, particularly interesting that the LRH phosphatase studied here was markedly stimulated in the presence of these exogenous polyribonucleotides. It is noteworthy that this activity was also stimulated by DOC, and thus may exist in a partially latent state in the envelope.

The isolation procedure for the pore complex-lamina fraction resulted in complete removal of phosphatase activity towards endogenous proteins and PV (not shown). However, the specific activity towards LRH was fully retained (as compared to that in intact envelope) in this subfraction (Table 3). Thus, it may be concluded that phosphatases active towards PV and endogenous pore-lamina proteins are either dissociated from this fraction (or inactivated) during its isolation or are not intrinsic to it.

CONCLUSION

The foregoing studies, utilizing multiple phosphoprotein substrates, suggest the presence of distinct multiple protein kinase and phosphatase enzymes in the nuclear envelope. These enzyme activities appear to be a general feature of the envelope since their existence has been demonstrated in envelopes derived from a number of different tissues (Steer et al. 1979a, b; Lam, Kasper 1979; Steer et al. 1980; Agutter et al. 1979a; McDonald, Agutter 1980; Clawson et al. 1980a, b; Kletzien 1981; Veneziale et al. 1981). They are unlikely to result from chromatin or ribosomal contamination for the following reasons: 1) nuclear envelope (and especially pore complex-lamina fraction) had minimal DNA or RNA; 2) envelope-associated chromatin has been identified as heterochromatin (Franke, Scheer 1974) which possesses low protein kinase activity (Ahmed et al. 1979; Norvitch et al. 1980); 3) extraction of envelopes with 2 M NaCl during isolation of the pore complex-lamina fraction should remove most of the phosphoproteins of chromatin origin; 4) the autoradiographic profile of ^{32}P-labeled envelopes showed essentially no activity in the region of the gel where highly phosphorylated chromatin proteins would appear (Ahmed et al. 1980); 5) labeling of nuclear envelope with ^{32}P followed a distinctly different pattern from that observed for microsomal preparations (Lam, Kasper 1979); 6) EDTA treatment of the partially purified envelope pellet affected the removal of ribosomes; and 7) chromatin contains minimal phosphatase activity and most of the nuclear phosphatase activity is extracted at low salt concentrations (Wilson et al. 1980).

The protein phosphatases and kinases associated with nuclear envelope share some characteristics which were observed for the respective enzymes in previous studies on total nuclei or chromatin (Ahmed, Wilson 1978; Ahmed et al. 1979, 1981; Goueli et al. 1980). This seems reasonable because of the presence of envelope components in such preparations. Certain distinct features of the envelope-associated phosphatase reactions e.g., stimulation by EDTA and lack of effect of ammonium molybdate, as well as stimulation by DOC and poly-A and poly-G, were not seen in studies of total nuclear-associated phosphatases.

A more detailed analysis shall be required to firmly establish the identity and relative topographic localization of the envelope-associated kinases and phosphatases, and their native substrates. Such knowledge should help to establish the possible physiological functions of these reactions. In this regard, some speculation can be presently made. Since a nucleoside triphosphatase has been implicated in the translocation of RNA across the envelope, it is conceivable that protein kinase and phosphatase reactions (by constituting an ATPase system) may participate in such a function. A number of correlative properties of the ATPase protein phosphatase described here have been noted, including similarities of pH optima, stimulation by polyribonucleotides, etc. Similar observations have also been made by McDonald and Agutter (1980).

Phosphorylation/dephosphorylation of envelope proteins may also influence the behavior of the envelope during the cell cycle and cellular growth. It has been reported that phosphorylation of NE proteins in proliferating cells in culture is markedly elevated as compared to that which occurs in quiescent cells (Kletzien 1981). In NE of seminal vesicle epithelium derived from androgen-deprived animals, there was only an equivocal indication of a decline in phosphorylation of envelope proteins (Veneziale et al. 1981); this, however, needs to be explored further. It shall also be of interest to determine if envelope phosphoproteins are involved in translocation of steroid-receptor complexes across the envelope. Recent studies have reported binding of steroid to the NE (Lefebvre, Novosad 1980; Barrack, Coffey 1980; Smith, von Holt 1981), however, the possible physiological significance of these observations remains to be established.

ACKNOWLEDGMENT

Original investigations reported in this article were supported in part by Veterans Administration Medical Research Fund, and by PHS grant CA 15062 awarded by the National Cancer Institute, DHHS.

Aaronson RP, Blobel G (1975). Isolation of nuclear pore complexes in association with a lamina. Proc Natl Acad Sci USA 72 (3):1007.
Agutter PS, Cockrill JR, Lavine JE, McCaldin B, Sim RB (1979a). Properties of mammalian nuclear envelope nucleoside triphosphatase. Biochem J 162:671.
Agutter PS, Harris JR, Stevenson I (1977). Ribonucleic acid stimulation of mammalian liver nuclear-envelope nucleoside triphosphatase. Biochem J 162:671.
Agutter PS, McCaldin B, McArdle HJ (1979b). Importance of mammalian nuclear-envelope nucleoside triphosphatase in nucleo-cytoplasmic transport of ribonucleoproteins. Biochem J 182:811.

Ahmed K (1975). Phosphoprotein metabolism in primary and accessory sex tissue. In Thomas JA, Singhal RL (eds): "Molecular Mechanisms of Gonadal Hormone Action", Vol 1, Advances in Sex Hormone Research, Baltimore: The University Park Press, p 129.

Ahmed K, Davis AT, Goueli SA, Wilson MJ (1980). Phosphorylation of a nonhistone protein fraction which co-extracts with the high-mobility-group proteins of chromatin. Biochem Biophys Res Commun 96:326.

Ahmed K, Wilson MJ (1978). Chromatin controls in the prostate. In Busch H (ed): "The Cell Nucleus", Vol VI, Chromatin, Part C, New York: Academic Press, p 409.

Ahmed K, Wilson MJ, Goueli SA (1981). Biochemistry of protein kinase reactions in the prostate in relation to androgen action. In Murphy GP, Sandbert A, Karr J (eds): "The Prostatic Cell: Structure and Function", Part A, Morphological, Secretory and Biochemical Aspects. New York: Alan R. Liss, Inc., p 55.

Ahmed K, Wilson MJ, Goueli SA, Norvitch ME (1979). Testosterone effects on the prostatic nucleus. In Busch M, Daskal Y, Crooke S (eds): "Effects of Drugs on the Cell Nucleus", New York: Academic Press, p 419.

Barrack ER, Coffey DS (1980). The specific binding of estrogens and androgens to the nuclear matrix of sex hormone responsive tissues. J Biol Chem 255:7265.

Clawson GA, Woo CH, Smuckler EA (1980a). Independent responses of nucleoside triphosphatase and protein kinase activities in nuclear envelope following thioacetamide treatment. Biochem Biophys Res Commun 95:1200.

Clawson GA, Woo CH, Smuckler EA (1980b). Polypeptide composition of nuclear envelope following thioacetamide-induced nuclear swelling. Biochem Biophys Res Commun 96:370.

Dwyer N, Blobel G (1976). A modified procedure for the isolation of a pore complex-lamina fraction from rat liver nuclei. J Cell Biol 70:581.

Franke WW, Scheer U (1974). Structures and functions of the nuclear envelope. In Busch H (ed): "The Cell Nucleus", New York: Academic Press, p 219.

Greengard P (1978). Phosphorylated proteins as physiological effectors. Science 199:146.

Harris JR (1978). The biochemistry and ultrastructure of the nuclear envelope. Biochim Biophys Acta 515:55.

Judah JD, Ahmed K (1964). The biochemistry of sodium transport. Biol Rev 39:160.

Kasper CB (1974). Chemical and biochemical properties of the nuclear envelope. In Busch H (ed): "The Cell Nucleus", New York: Academic Press, p 349.

Kay RR, Fraser D, Johnston IR (1972). A method for the rapid isolation of nuclear membranes from rat liver. Eur J Biochem 30:145.

Kletzien RF (1981). Nuclear membrane-associated protein kinase and substrates. Biochem J 196:853.

Lam KS, Kasper CB (1979). Selective phosphorylation of a nuclear envelope polypeptide by an endogenous protein kinase. Biochemistry 18:307.

Lefebvre YA, Novosad Z (1980). Binding of androgens to a nuclear-envelope fraction from the rat ventral prostate. Biochem J 186:641.

McDonald JR, Agutter PS (1980). The relationship between polyribonucleotide binding and the phosphorylation and dephosphorylation of nuclear envelope protein. FEBS Lett 116:145.

Norvitch ME, Wilson MJ, Ahmed K (1980). Distribution of protein phosphokinases and chromosomal phosphoproteins in heterochromatin and euchromatin of rat ventral prostate. Eur J Cell Biol 22:78.

Rubin CS, Rosen OM (1975). Protein phosphorylation. Ann Rev Biochem 44:831.

Smith P, von Holt C (1981). Interaction of the activated cytoplasmic glucocorticoid hormone receptor complex with the nuclear envelope. Biochemistry 20:2900.

Steer RC, Goueli SA, Wilson MJ, Ahmed K (1980). Cobalt-stimulated protein phosphokinase activity of the pore complex-lamina fraction from rat liver nuclear envelope. Biochem Biophys Res Commun 92:919.

Steer RC, Wilson MJ, Ahmed K (1979a). Protein phosphokinase activity of rat liver nuclear membrane. Exp Cell Res 119:403.

Steer RC, Wilson MJ, Ahmed K (1979b). Phosphoprotein phosphatase activity of rat liver nuclear membrane. Biochem Biophys Res Commun 89:1082.

Veneziale CM, Utz ME, Steer RC, Wilson MJ, Ahmed K (1981). Nuclear envelope of the seminal vesicle epithelium. Biochem J 198:259.

Wilson MJ, Goueli SA, Ahmed K (1980). Partial purification of nuclear protein phosphatases of rat ventral prostate. Proceedings of 6th Int. Congr. Endocrinology, Melbourne, Australia, p 545.

PHOSPHOLIPID METABOLISM WITH RESPECT TO NUCLEAR ENVELOPE REFORMATION IN LATE MITOSIS IN HeLa S_3 CELLS

S.M. Henry and L.D. Hodge

Department of Cell and Molecular Biology
Medical College of Georgia
Augusta, Georgia 30912

INTRODUCTION: LATE MITOSIS AND LABELING OF PHOSPHOLIPID

The major ultrastructural events which occur during re-establishment of the interphase state as metaphase cells traverse through anaphase and telophase and into early G_1 have been described in detail and are widely known. The biochemical aspects of the re-establishment, particularly the membranous events, and especially the nuclear envelope (NE) events, have received considerably less attention. Obviously knowledge of the membrane changes at the biochemical level is necessary for a clear understanding of these processes per se, their control, if any, and their possible interrelationship with regard to other ultrastructural events and the re-establishment of interphase function. Thus, our goal is to examine phospholipid metabolism in late mitotic stages to determine whether or not membrane changes that occur during late mitosis can be correlated with specific metabolic events.

To study the biochemical events of NE reformation, a tightly synchronized population of metaphase cells, as well as an easily identified specific membranous component(s), is essential. To obtain this population, we selectively detach metaphase cells from monolayer cultures of double thymidine-arrested HeLa S_3 cells, as previously described (Hodge et al. 1969). The mitotic index of the detached population is routinely 90 to 95% with 80 to 90% in the metaphase stage of mitosis. When these cells are subsequently incubated at 37°C, essentially all of the cells traverse to early G_1 such that by 90 min only interphase cells are present. Incubation for lesser times yields populations of predominantly late anaphase to early telophase, or late telophase to early G_1 cells. Thus, tightly synchronized populations in late mitotic stages can be obtained.

An easily identified, specific (and noncontroversial) membranous component are the phospholipids. Compounds such as radioactive orthophosphate and glycerol can be used to label all classes of phospholipids, except sphingomyelin in the case of the glycerol precursor. On the other hand, radioactive choline, inositol and the methyl group of methionine can, for example, be used as labels for specific classes of phospholipids.

The nature of the questions concerning phospholipid metabolism which can be asked depend in part on the types of labeling regimens which can be performed. With late mitotic cells, as shown in Figure 1, four such standard procedures can be carried out. These are: 1) continuous labeling with radioactive glycerol, choline, orthophosphate and methionine during the time required for metaphase cells to reach G_1 (panel A); 2) pulse labeling with radioactive glycerol or choline for 10 min at various intervals during the metaphase to G_1 transition (panel B); 3) pulse-chase labeling in which the incorporation of radioactive glycerol can be terminated by an excess of unlabeled precursor (panel C); and finally 4) prelabeling with glycerol for 22 hr preceding the mitotic detachment procedure (panel C).

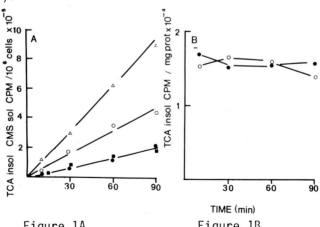

Figure 1A   Figure 1B

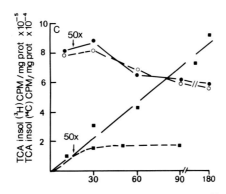

Figure 1C [Time in minutes]

Figure 1. Phospholipid labeling procedures with late mitotic cells.

Panel A: Metaphase cells at a concentration of 1 x 10^6/ml were incubated at 37°C in the presence of either [^{3}H]glycerol (10 µCi/ml), [^{3}H]choline (10 µCi), [^{3}H]methylmethionine (25 µCi/ml), or [^{32}P]orthophosphate (0.15 mCi/ml). At the indicated times, aliquots of each culture were washed and analyzed for incorporation of radioactivity. Glycerol and choline incorporation were scored as trichloracetic acid (TCA)-insoluble radioactivity. Methyl group and phosphate incorporation were scored as chloroform-methanol-soluble (CMS) radioactivity (Abdel-Latif et al. 1977). (■——■, glycerol incorporation; ●———●, choline incorporation; △————△, orthophosphate incorporation; o————o, methyl group incorporation).

Panel B: Populations of late mitotic cells were pulse-labeled for 10 min with radioactive glycerol or choline as described above. TCA-insoluble radioactivity was determined. (o————o, glycerol incorporation; ●———●, choline incorporation).

Panel C: For a pulse-chase, mitotic cells were exposed to [^{14}C]glycerol (2 µCi/ml). After 15 min, the culture was divided into equal aliquots and a 50-fold excess of unlabeled glycerol was added to one. TCA-insoluble radioactivity and protein were determined at the times indicated (■——■, [^{14}C]glycerol incorporation; ■---■, [^{14}C]glycerol incorporation in the presence of excess unlabeled glycerol). For prelabeling mitotic cells, cells were labeled with glycerol (5 µCi/ml) during synchronization. Mitotic cells were then collected, divided into two cultures and incubated at 37°C. After 15 min, a 50-fold excess of nonradioactive glycerol was added to one culture. At the times indicated, TCA-insoluble radioactivity and protein were determined. (●——●, presence of excess unlabeled glycerol; o———o, absence of unlabeled glycerol).

In these experiments we investigated incorporation into whole cells because lipid biosynthetic reactions, although preferentially associated with the endoplasmic reticulum, are distributed throughout the cell membranous system (Morre et al. 1979), and also because isolation of a NE obviously was not possible from metaphase cells.

In addition to providing information upon which experiments can be based, these labeling methodologies provide some information about the phospholipid biosynthetic activity of late mitotic cells. As demonstrated by the continuous incorporation of the four different precursors (Fig. 1, panel A), late mitotic cells are engaged in the synthesis of phospholipids. As indicated by the pulse labeling conditions (panel B), all stages of late mitotic cells appear to be equally active in the synthesis of phospholipids. Finally, as expected (panel C), the vast majority of phospholipid prelabel survives division, although the data do suggest that a fraction of this label is apparently discarded. This recovery is apparently not influenced by a 50-fold excess of unlabeled glycerol. The nature of the surviving phospholipid fraction could be examined and compared to the synthesis of new phospholipids. Such comparisons would not be only a means by which to study the potential role of phospholipid metabolism in the reformation of the NE but would also be a means by which to study the role of pre-existing phospholipids in this process.

ANALYSIS OF PHOSPHOLIPID METABOLISM BY RAPID LABELING

As detailed in panel A above, we have begun to analyze phospholipid metabolism by means of a rapid labeling procedure. The length of the labeling time is significantly shorter than the half-life of the major classes of phospholipids (for a review, see Van den Bosch 1980). The level of activity in late mitotic cells has been assessed and compared to the level of activity at two subsequent times. We chose for such a comparison two G_1 stages, one at 4 hr postmetaphase when the cells are in mid G_1 and the second at 6.5 hr after metaphase when the cells are in late G_1. A few such comparisons appear in the literature, however, it is worth noting that these data were obtained with populations of cells which were at a maximum only 30 to 40% mitotic (eg., Conner et al. 1980). Thus it is not possible to conclude whether the observed synthesis occurred in late G_2 and/or mitosis and/or G_1. An additional complication in the interpretation of these published observations results from the use of the long labeling times, eg., 4 hr (Hirschberg et al. 1974).

Late mitotic cells, mid G_1 and late G_1 cells were exposed to the four phospholipid precursors for 90 min and the radioactivity recovered in the TCA-water-soluble and CMS fractions was determined (Table 1). The CMS data indicate that the level of incorporation of three of the lipid precursors, orthophosphate, glycerol, and choline, is higher by nearly 2-fold during the M to G_1 transition as compared to mid or late G_1 cells.

In contrast, no difference was noted in the incorporation of methyl groups. The amount of radioactivity for each label recovered in the soluble fractions in both mitotic and post-mitotic cells was essentially the same. It appears, therefore, that the differences noted in the recovery of radioactivity in the phospholipids should not be attributed to differences in uptake of the precursor or synthesis of water-soluble metabolites, such as phosphorylcholine. Thus some pathways in the synthesis of phospholipids appear to be augmented in late mitotic cells, whereas, at least one, the phosphatidylethanolamine (PE) to phosphatidylcholine (PC) pathway which is assessed by methyl group incorporation is not augmented. These observations raise questions concerning the importance of the augmented and nonaugmented pathways to the reestablishment of the interphase state which will be dealt with to some extent below.

An analysis of the labeled phospholipids by a two-dimensional chromatographic separation, which can fraction-

Cell Stage	Radioactive Label			
	orthophosphate	glycerol	choline	methionine
M + G_1 TCA-H_2O S (cpm/10^6 cells)	8.3×10^6 (100%)	5.6×10^3 (100%)	3.1×10^5 (100%)	4.2×10^4 (100%)
CMS/Pi (cpm/µmole)	1.6×10^7 (100%)	4.8×10^6 (100%)	2.5×10^6 (100%)	1.8×10^6 (100%)
Mid G_1 TCA-H_2O S (cpm/10^6 cells)	8.3×10^6 (99.6%)	5.3×10^3 (94.8%)	3.5×10^5 (113%)	4.6×10^4 (109.5%)
CMS/Pi (cpm/µmole)	1.0×10^7 (69.4%)	2.7×10^6 (56.2%)	1.6×10^6 (64.0%)	1.7×10^6 (95%)
Late G_1 TCA-H_2O S (cpm/10^6 cells)	8.1×10^6 (97.5%)	5.4×10^3 (96.2)	3.6×10^5 (117%)	4.7×10^4 (111%)
CMS/Pi (cpm/µmole)	1.0×10^7 (68.3%)	2.5×10^6 (52%)	1.5×10^6 (60%)	1.6×10^6 (90%)

Table 1. Incorporation of lipid precursors in M to G_1 cells, mid G_1 cells and late G_1 cells. HeLa cells were synchronized and populations in metaphase, mid (4 hr) and late (6.5 hr) G_1 were obtained. Each population was exposed to radioactive lipid precursor as described in Figure 1, panel A. At the end of a 90-min period, the TCA-water-soluble (TCA-H_2O-S) and CMS fractions were prepared, and the amount of radioactivity and phospholipid phosphate (Pi) were determined (Abdel-Latif et al. 1977).

ate total phospholipids into 15 classes (Abdel-Latif et al. 1977), has provided additional information (Table 2).

Because the number of metaphase cells which can be obtained by selective detachment is relatively small, it was necessary to add carrier lipids obtained from log phase HeLa cells to locate the phospholipids. By this means, we were routinely able to identify a dozen phospholipid classes including all the major ones. Needless to say, this prevented us from determining the radioactivity per unit of Pi for each phospholipid class, and in routine experiments we have therefore, made comparisons based on the relative distribution of radioactivity among these classes. After orthophosphate and glycerol labeling, radioactivity was recovered in all the separated phospholipids, including those extracted from the late stages of mitosis. Several differences in the relative distribution of these labels were noted. The most notable was the relative increased recovery of label in the choline-containing phospholipids during the M to G_1 transition, as noted in Table 2, below. After choline and methionine labeling, the recovery of label indicated that the

Radioactive Label	Cell Stage	Classes of Phospholipid		
		LPC	SpH	PC
Orthophosphate	M → G_1	8.3 ± .56	1.01 ± .25	21.85 ± 1.06
	Mid G_1	6.1 ± .99	1.65 ± .21	13.55 ± .21
	Late G_1	5.8 ± .56	<1	9.4 ± .28
Glycerol	M → G_1	10.65 ± 1.77	<1	56.8 ± 2.26
	Mid G_1	7.15 ± 1.6	<1	42.95 ± 2.4
	Late G_1	7.7 ± .91	<1	35.4 ± 2.4
Choline	M → G_1	15.67 ± 3.64	1.33 ± .21	82.33 ± 3.76
	Mid G_1	13.9 ± 4.2	1.5 ± .61	83.9 ± 4.2
	Late G_1	16.25 ± 2.05	1.55 ± .92	81.3 ± 2.83
Methionine	M → G_1	18.17 ± 1.63	2.5 ± .32	72.8 ± .47
	Mid G_1	14.25 ± 1.20	1.4 ± .71	74.1 ± 3.39
	Late G_1	12.8 ± 3.39	1.7 ± .14	78.45 ± 8.7

Table 2. Relative distribution of radioactivity in three classes of phospholipids. After labeling M to G_1 cells, mid G_1 cells and late G_1 cells for 90 min, as described in the legend of Figure 1 (panel A), a CMS fraction was prepared and was analyzed by two-dimensional thin-layer chromatography (Abdel-Latif et al. 1977). The relative recovery of radioactivity in all the phospholipids was determined. The data shown express the relative recovery of radioactivity in PC, lysophosphatidylcholine (LPC) and sphingomyelin (SPH). Values represent the average ± SD of three separate experiments for each cell stage and label.

relative synthesis of these choline-containing derivatives was similar in both mitotic and post-mitotic times.

Because PC is one of the most abundant classes of phospholipids in mammalian cells, it was possible to locate this class on chromatographic plates without the addition of carrier lipid; therefore, we have extended our observation concerning the incorporation of label into this phospholipid. By quantifying the incorporation of label per micromole of recovered PC, the data indicated that there was a 2- to 3-fold increase in the level of synthesis during the late mitosis as compared to subsequent G_1.

Radioactive Label	Cell Cycle Stage		Ratio $M \rightarrow G_1$/ Late G_1
	$M \rightarrow G$ (cpm/μmole Pi)	Late G_1 (cpm/μmole Pi)	
Orthophosphate	2.1×10^6	0.72×10^6	2.9
Glycerol	1.8×10^6	0.86×10^6	2.1

Table 3. Specific activities of PC in M to G_1 and late G_1 cells. M to G_1 and late G_1 cells were labeled with either radioactive orthophosphate or glycerol as described in the legend of Figure 1. Phospholipids were extracted and were separated by two-dimensional chromatography. The PC spots were eluted, and radioactivity and Pi were determined.

To visualize the intracellular site(s) of phospholipid precursor incorporation, we examined individual cells by electron microscope autoradiography. Cells synchronized in metaphase were suspended in choline-deficient medium containing tritiated choline (100 μCi/ml) and permitted to progress to late mitotic stages. Typical cells are shown in Figure 2.

By early telophase, obvious sites of incorporation were seen at the periphery of the newly reformed nucleus before significant chromosome decondensation had begun,

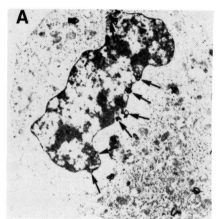
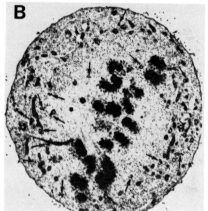
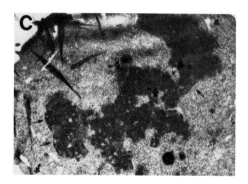

Figure 2. Visualization of choline incorporation in late mitotic cells. Metaphase cells were suspended in medium containing 0.01 x choline and [^{3}H]choline at a concentration of 100 µCi/ml as described in the legend of Figure 1. Incubation was continued for 40 min at 37°C in order to observe late stages of mitosis. After fixation and embedding, 1500 Å sections were autoradiographed as described by Simmons et al. 1973. Panel A is a section through a typical telophase cell, 10,600X. Panel B is a section through a typical metaphase cell, 10,600X. Panel C is a section through a typical anaphase cell at higher magnification to detail partial reformation of the nuclear envelope, 26,000X. Arrows indicate the location of grains.

and before nucleolar reformation (Fig. 2A). The autoradiographic grains in metaphase cells, in contrast, were distributed throughout the cytoplasm toward the cell periphery, and at some distance from the electron dense chromosomes (Fig. 2B). In metaphase cells, the grains appeared to be associated with known cytoplasmic structures, such as mitochondria and vesicles. In late anaphase, before complete envelope reformation was obvious, only small numbers of grains were seen at the site of the fusing chromosome mass (Fig. 2C). It is of interest that these grains were observed both at sites where NE was apparent and at sites where none was obvious. Based on the observation in anaphase, it is tempting to speculate that the incorporation of PC precedes NE reformation in late mitosis, although this is difficult to assess.

In order to quantify these observations, we have counted total cytoplasmic grains, as well as grains within three half-distances of the chromosomal or nuclear periphery and the plasma membrane in anaphase, telophase and G_1 cells (Table 4). At least 85% of the grains associated with the radioactivity in a given structure should lie within this distance (Salpeter et al. 1969). Consistent with the previous observation, we find, at least by telophase, a signi-

Cellular Compartment	Stage in Cell Cycle		
	Anaphase %	Telophase %	Early G_1 %
Nuclear Envelope	10.5 (84)	22.6 (227)	30.6 (231)
Plasma membrane	24.7 (198)	24.0 (243)	26.8 (202)
Cytoplasm (other membraneous structure)	64.8 (519)	53.2 (536)	42.6 (322)

Table 4. Intracellular distribution of choline label. The intracellular distribution of grains was determined on autoradiographs of 30 anaphase, 30 telophase and 30 early G_1 cells. Electron micrographs were printed at 12,500X. Numbers in parentheses are the total number of grains observed over that cellular compartment or within three half-distances of the cellular structure.

ficant percentage of grains associated with the NE. This represents nearly twice the percentage of grains observed at the periphery of the nuclear mass in late anaphase. Significant numbers of grains were also seen associated with the plasma membrane, as well as with cytoplasmic membranous structures at each late mitotic stage.

Based on cell fractionation, we calculate that approximately 18% of the total cell phospholipid is associated with the NE in early G_1 cells. Thus, an estimate of the labeling index for the NE in these cells would be 1.7 (30.6%/18%) and for the plasma membrane and cytoplasm 1.09 (89.4%/82%). These values are consistent with a significant synthesis of NE phospholipids, as compared to other membranous structures, during late mitosis.

PC AND NE REFORMATION

So far, we have presented data indicating: 1) that there is significant synthesis of several classes of phos-

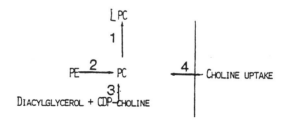

Figure 3. PC metabolism. The figure depicts known pathways of PC metabolism and their inhibitors:

1. Mepacrine - inhibits phospholipase A_2 (Hirata et al. 1979)
2. 3-Deazaadenosine - inhibits methylation of PE (Hirata, Axelrod 1980)
3. Centrophenoxine - inhibits choline phosphotransferase (Parthasarathy et al. 1978)
4. Hemicholinium - inhibits transport into cell (Abdel-Latif, Smith, 1972)

pholipids in late mitotic cells, 2) that pathways for the de novo synthesis of PC appear to be augmented in late mitosis, 3) that the synthesis of PC by methylation of PE occurs and is not augmented, and 4) that the NE is a site of significant PC incorporation in late mitotic cells. These observations suggest, for one thing, that continued study of PC metabolism during late mitosis could offer further insights into NE reformation and the relationship of this event to other events in the re-establishment of interphase function.

With regard to this idea, one approach to the study of PC metabolism is the use of inhibitors. The known metabolic pathways for this lipid are depicted schematically in Figure 3 along with several inhibitors.

Recently we have begun studies using one such compound, 3-deazaadenosine, which preferentially inhibits the PE to PC conversion by an effect on S-adenosylmethionine metabolism. Addition of 1 mM deazaadenosine to metaphase cells results in an altered ultrastructural appearance of nucleoli and the NE in early G_1 cells, as determined by electron microscopic examination (Fig. 4).

There appears to be only partial nucleolar reformation and "improper" NE reformation in all cells examined, as judged by their altered morphological appearance. The range of NE effects includes loss of ultrastructural detail and/or apparent absence of envelope, with or without interspersed envelope of normal morphology. These effects seem to occur in the absence of an effect on cytoplasmic events such as polysome reformation and division in early G_1.

When such cells were examined biochemically using the inhibitor and radioactive methionine, we found approximately 80% inhibition of the incorporation of methyl groups into the CMS fraction. When the phospholipids from such treated cells were separated by two-dimensional chromatography, the radioactivity recovered in PC was decreased by about 75%; presumably the remainder of the radioactivity was distributed in unidentified intermediates in the PE to PC conversion (Hirata et al. 1979). The specificity of the inhibitor in our system is indicated by our observation that choline and glycerol incorporation in late mitotic cells in the presence of deazaadenosine was at control levels. These data suggest that the maintenance of at least nonaugmented path-

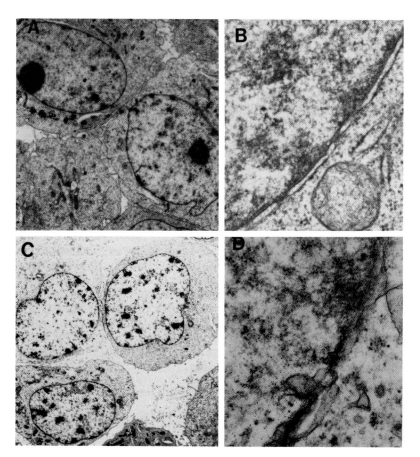

Figure 4. Ultrastructural appearance of early G_1 cells treated with deazaadenosine at metaphase. Panel A depicts a G_1 cell from untreated cultures showing complete nucleolar and NE reformation, 6,000X. Panel B is a high-power view showing the normal morphological appearance of the NE which is characterized by pore structures, a well-defined intermembranous space, and a uniform appearance to the outer membrane, 60,000X. Panel C depicts a G_1 cell from a treated culture showing "incomplete" nucleolar reformation and an altered NE, 6,000X. Panel D is a high-power view showing the abnormal morphological appearance of the NE in which the ultrastructural details noted in Panel B are distorted, 60,000X.

Kennedy EP, Weiss SB (1956). The function of cytidine coenzymes in the biosynthesis of phospholipids. J Biol Chem 222:193

Morre DJ, Kartenbeck J, Franke WW (1979). Membrane flow and interconversions among endomembranes. Biochim Biophys Acta 559:71.

Parthasarathy S, Cady RK, Kravshaar DS, Sladek NE, Baumann WJ (1978). Inhibition of diacylglycerol: CDP choline phosphotransferase activity by dimethylaminoethyl p-chlorophenoxyacetate. Lipids 13:161.

Salpeter MM, Bachmann L, Salpeter EE (1969). Resolution in electron microscope radioautography. J Cell Biol 41:1.

Simmons T, Heywood P, Hodge LD (1973). Nuclear envelope-associated resumption of RNA metabolism in late mitosis of HeLa cells. J Cell Biol 59:150.

Van den Bosch H (1974). Phosphoglyceride metabolism. Annual Rev Biochem 43:243.

Van den Bosch H (1980). Intracellular phospholipases A. Biochim Biophys Acta 604:191.

THE UPTAKE OF NUCLEAR PROTEINS IN OOCYTES

Carl M. Feldherr and Joseph A. Ogburn

Department of Anatomy
University of Florida College of Medicine
Gainesville, Florida 32610

INTRODUCTION

The experiments reported here relate to the nucleo-cytoplasmic exchanges of endogenous proteins and were designed to study the possible mechanisms which regulate the intracellular distribution of these molecules. <u>Xenopus</u> oocytes were selected as the experimental material for these investigations because of their size, the rapidity with which the germinal vesicles can be isolated and their ability to tolerate microinjection procedures.

In this paper we will present evidence supporting the view that exchanges of endogenous proteins between the nucleus and cytoplasm are controlled by selective processes. To put these findings in perspective it is useful to consider the results of previous studies on oocytes in which the nuclear uptake of exogenous tracers was investigated. Presumably these substances enter the nucleus by nonselective mechanisms which reflect the inherent limitations imposed by the envelope and, perhaps, other cell components. Intact oocytes have been microinjected with a number of different substances including tritiated dextrans (Paine et al. 1975), a variety of $[^{125}I]$-labeled and fluorescein-labeled proteins (Bonner 1975a; Paine, Feldherr 1972), and colloidal particles (Feldherr 1969). All of the data obtained are consistent with the view that exogenous macromolecules enter oocyte nuclei by passive diffusion through the pores of the nuclear envelope; however, the movement of these molecules is restricted, and the degree of restriction appears to be directly related to molecular size. It is estimated that molecules with diameters larger than 90 Å are unable to diffuse through the pores (Paine et al. 1975).

NUCLEAR UPTAKE OF ENDOGENOUS PROTEINS

Initially, it was necessary to determine if the nuclear uptake rates of different size endogenous proteins were comparable to those observed for exogenous substances (Feldherr 1975). To accomplish this, Xenopus oocytes were labeled with tritiated amino acids and enucleated at intervals varying from 30 min to 6 hr. Nucleoplasmic and cytoplasmic samples were then run on one-dimensional sodium dodecyl sulfate (SDS)-polyacrylamide gels. The relative amounts of different size polypeptides present in the two regions of the cell were determined by measured radioactivity in gel slices.

The major finding in this investigation was that newly synthesized polypeptides, with molecular weights ranging from 94,000 to 150,000, were approximately twice as concentrated in the nucleus after only 3 hr. These rates of accumulation are significantly greater than would be expected for exogenous molecules of comparable size. Although this study demonstrated a difference in accumulation rates, it was not possible to decide if the results were due to the selective uptake of endogenous proteins by the nuclear envelope or other processes, such as nucleoplasmic building.

EXPERIMENTAL ALTERATIONS OF NUCLEAR ENVELOPE PERMEABILITY

To evaluate the function of the envelope in regulating the distribution of endogeneous proteins, its permeability characteristics were altered by puncturing the oocytes in the region of the germinal vesicles with fine glass needles (Feldherr and Pomerantz 1978). It was first to show that the envelopes were, in fact, disrupted by the puncture procedure and that the nuclear permeability characteristics were altered. Using the electron microscope, large gaps were consistently observed in the envelopes of punctured, but not control, oocytes. Functional differences were demonstrated by studying the nuclear uptake rates of cytoplasmically injected [^{125}I]BSA. It was found that disruption of the envelope resulted in an eleven-fold increase in the nuclear incorporation of this protein.

After finding that the ability of the envelope to act as a diffusion barrier could be experimentally altered, the effect of puncturing on the distribution and nuclear uptake of endogenous proteins was investigated. A group of oocytes was labeled with [^{3}H]leucine, after which half of the cells were punctured. The nuclei of both the punctured and nonpunctured cells were isolated either 30 min or 2 1/2 hr later, and run on one-dimensional SDS-polyacrylamide gels. No differences were detected in either the Coomassie Blue staining patterns or the rates at which different size classes of polypeptide entered the nucleoplasm. The results suggest that the nuclear envelope does not have a major role in regulating either the nucleocytoplasmic distribution of endogenous proteins or their rates of exchange. Apparently, the accumulation of specific proteins in the nucleus is due to selective binding. Because of the limitations of one-dimensional gels, these conclusions apply mainly to groups of polypeptides having similar molecular weights.

To extend the above studies and determine the effect of the envelope on the distribution of individual or, at least small, groups of polypeptides, experiments were conducted using two-dimensional gels (Feldherr, Ogburn 1980). Figure 1 shows a typical gel pattern obtained for oocyte nuclear proteins using the methods described by O'Farrell and O'Farrell (1977).

The following procedures were used to determine if disruption of the envelope causes qualitative changes in the protein composition of nucleus: Oocytes were labeled with [^{3}H]leucine, immediately punctured and enucleated 2 1/2 hr later. Two-dimensional gels were then prepared and fluorographed. When the results were compared to those obtained for nonpunctured cells, variations were found in only about 4% of the 300 nuclear polypeptides that could be identified. In one experiment, for example, nine polypeptides that were present in the punctured nuclei were not observed in the controls, whereas four other polypeptides were found only in the controls. Furthermore, different polypeptides varied in different experiments. These results confirmed our earlier findings that the nucleocytoplasmic distribution of proteins is not controlled primarily by the envelope.

Quantative studies, using two-dimensional gels, were conducted to investigate the effect of puncturing on the nuclear accumulation rates of specific proteins. Five polypeptides were selected for analysis, including N1, N2, B3, B4 and actin (see Fig. 1). All of these molecules incorporated sufficient amounts of labeled amino acids to permit determinations of uptake rates. In addition, they represented a diversity of sizes (from 43,000 to 110,000 molecular weight) and intracellular distribution patterns (N1 and N2 accumulate in the nucleus, whereas B3, B4 and actin are present in equivalent concentrations in the nucleus and cytoplasm). Analysis of the nuclear accumulation rates of these polypeptides was carried out as follows: The oocytes that were to be punctured, maintained as controls, or used to establish equilibration times, were labeled with [^{3}H]leucine and enucleated after the appropriate time period. In all instances the [^{3}H]-labeled nuclei

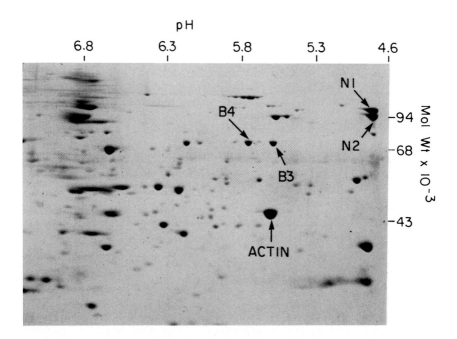

Figure 1. A two-dimensional SDS-polyacrylamide gel, stained with Coomassie Blue, showing the normal distribution of nuclear polypeptides. The polypeptides that were analyzed in the quantitative studies are labeled (from Feldherr, Ogburn 1980).

were electrophoresed along with [^{35}S]-labeled nuclei, which served as internal standards. To obtain the standards, oocytes were labeled with [^{35}S]methionine and enucleated immediately afterwards. When electrophoresis was completed, the relevant spots were cut out of the gel and the ^{3}H:^{35}S ratios were determined.

Prior to studying the effect of puncturing on the uptake of nuclear proteins, it was necessary to find the equilibration times for the polypeptides that were to be analyzed. Using the double-labeling procedure described above, the following values were obtained: 1-3 hr for B3 and B4; 6-9 hr for actin; 9-12 hr for N1 and N2.

With these data, experiments were designed in which punctured and nonpunctured oocytes were enucleated 1 or 2 hr after labeling with [^{3}H]leucine, that is, before equilibration was achieved. The nuclei were electrophoresed with ^{35}S standards, as described. The results are given in Table I. The experimental error was found to be ± 0.2.

Table 1

Effect of Puncturing on the Uptake of Nuclear Proteins

Protein	1 hr	2 hr*	2 hr	2 hr
N1	1.1	1.0	0.8	0.9
N2	1.3	1.2	0.9	0.9
B3	1.1	---	1.0	1.2
B4	1.2	---	1.1	1.2
Actin	1.8	2.2	1.4	1.5

*Three separate 2 hr experiments were performed.

The results are expressed as the ^{3}H:^{35}S puncture ratio divided by the ^{3}H:^{35}S control ratio. Approximately 20 experimental nuclei were used in each experiment. Data from Feldherr and Ogburn (1980).

Of the polypeptides that were analyzed, only actin accumulated more rapidly in punctured nuclei. This is indicated by the fact that, in all instances, the puncture:control ratios obtained for this protein were significantly greater than 1. The uptake of the remaining polypeptides was not affected, demonstrating that the envelope does not function as a rate-limiting barrier to these molecules, as is the case for exogenous tracers.

There are two general mechanisms which can explain the observed differences in the function of the envelope with regard to exogenous and endogenous exchanges. First, selective mechanisms might exist to facilitate the passage of specific endogenous proteins either through the pores or across the membranes of the envelope. Second, the rates at which endogenous molecules accumulate in the nucleus might be regulated by nucleoplasmic or cytoplasmic binding.

NUCLEAR BINDING

It has been reported that N1 and N2 are approximately 120 times more concentrated in the nucleus than in the cytoplasm (DeRobertis et al. 1978). Since the equilibrium distribution of these molecules is not altered by disruption of the envelope (Feldherr, Ogburn 1980) it can be concluded that they are bound within the nucleoplasm. Furthermore, pulse-labeled molecules, which enter the nucleus, remain concentrated for at least 24 hr, indicating that they bind to high affinity sites. Due to these factors, N1 and N2 are especially useful for studying the possible role of nuclear binding in regulating protein uptake.

Since high affinity sites are apparently involved, binding would be the rate-limiting factor for the accumulation of N1 and N2 if 1) there is a continuous increase in the number of available binding sites, and 2) the sites are occupied as rapidly as they are produced. To test this hypothesis, experiments were designed to determine if excess binding sites normally exist for these two polypeptides. The presence of additional sites would argue against binding as a regulatory factor, whereas an absence of such sites would support, but not prove, the hypothesis.

Basically, the experimental procedures involved the preparation of nuclear extracts from labeled oocytes and the microinjection of the extracts into either the nucleus or cytoplasm of intact cells. In this way the amounts of N1 and N2 in either region of the cell could be experimentally altered and their subsequent redistribution determined.

Nuclear extracts were prepared, with minor modifications, according to the procedures outlined by DeRobertis et al. (1978). Oocytes were labeled overnight with [^{3}H]leucine (1 mCi/2 ml of Ringer's solution), manually enucleated in intracellular medium containing PVP, homogenized, extracted for 2 hr and centrifuged at 100,000 x g for 15 min. It should be pointed out that B3, B4 and actin were also extracted, but they were not highly concentrated and could not be analyzed in subsequent injection experiments. There are several lines of evidence indicating that N1 and N2 are not altered by the extraction procedures. When compared to endogenous N1 and N2, the extracted molecules exhibited the same migration pattern on two-dimensional gels, and achieved the same equilibrium distribution when injected into the cytoplasm (approximately 75-80% of the total cellular N1 and N2 are present in the nucleus). The extracted molecules did take longer to reach equilibrium; however, this does not necessarily reflect a functional change since the delay corresponded to the time required for the injected polypeptides to diffuse throughout the cell.

The techniques used for nuclear injection were similar to those reported by Kressmann and Birnstiel (1980). The oocytes were centrifuged at a speed sufficient to move the nuclei to the surface of the animal pole, where they were injected with approximately 20 nl of extract. Knowing the number of nuclei extracted, the extracted efficiency and the equilibrium distribution, it was estimated that the amounts of N1 and N2 contained in 20 nl of extract were equivalent to approximately 20% of the normal cytoplasmic pools of these molecules.

The nuclei were isolated 2 hr after injection and run on two-dimensional gels along with [^{35}S]-labeled standards. A volume of extract, equal to the amount that was injected into the nuclei, was also electrophoresed together with [^{35}S]-labeled nuclei. The proportions of N1 and N2 retained by the injection nuclei were determined by comparing the ^{3}H:^{35}S ratios observed for the nuclei to those obtained for the extract. The results of these experiments are given in Table 2. The data show that essentially all of the injected N1 and N2 remain localized within the nuclei.

The retention of injected N1 and N2 can be explained in one of two ways; first, they could bind within the nucleoplasm or, second, their diffusion out of the nucleus could be restricted by the envelope. To distinguish between these possibilities, nuclear injections were performed on two groups of oocytes; the nuclei in one group were punctured, and the second group served as nonpunctured controls. Two hours later the nuclei were isolated and electrophoresed with ^{35}S standards. The results (see Table 2) showed that puncturing did not significantly affect the distribution of injected N1 and N2, indicating that retention of these molecules was due to binding. If N1 and N2 were not bound, but could freely diffuse, it is estimated that there would be a 70% loss of labeled protein from the nuclei during the 2-hr interval between injection and enucleation.

Table 2

Nuclear Injection Results

Experiment	Nuclei Extract		Puncture Nonpuncture	
	N1	N2	N1	N2
1	0.9	0.9	0.8	0.8
2	1.1	0.9	1.1	1.0
3	0.8	0.8		

The results for both sets of data are expressed as the ^{3}H:^{35}S ratios obtained for the first experimental group divided by the ratios obtained for the second group. See text for additional details. The number of nuclei analyzed in each experimental group varied from 9 to 18. The experimental error was ± 0.2.

It can be concluded from this study that the nuclei contain an excess number of binding sites for N1 and N2; therefore, it is unlikely that binding is rate-limiting for the accumulation of these specific polypeptides.

If these conclusions are correct, it should be possible to increase the total nuclear uptake of N1 and N2 by injecting extract into the cytoplasm. This is consistent with the finding that 12 hr after a 50-nl injection, the amount of extracted N1 and N2 incorporated by the nuclei was equivalent to approximately 40% of the endogenous N1 and N2 that is normally taken up during the same period. However, to demonstrate that there is a net increase in nuclear N1 and N2 it is necessary to show, in addition, that the uptake of endogenous protein is not affected by the extract. This was accomplished by injecting nonlabeled extract into oocytes which had been labeled with [^{3}H]leucine. As a control, labeled cells were injected with 50 nl of intracellular medium. The nuclei were isolated either 2 hr or 12 hr after injection and were analyzed on two-dimensional gels. It was found that neither the nuclear uptake rates nor the equilibrium distribution of endogenous N1 and N2 were affected by the extract. The results demonstrate that the nuclear uptake of these polypeptides is dependent on their cytoplasmic concentration, and provide further support for the view that nuclear binding is not rate-limiting. Unfortunately, the total binding capacity of the nucleoplasm could not be determined since we were unable to substantially increase the concentration of N1 and N2 in the nuclear extracts.

DIRECT PENETRATION OF THE NUCLEAR ENVELOPE

There is convincing evidence that proteins can be transported directly across the membranes of organelles such as the endoplasmic reticulum and mitochrondria (e.g., see review by Lusis and Swank 1980). It is appropriate, therefore, to consider whether or not similar mechanisms exist for the nuclear uptake of polypeptides.

Two possible mechanisms have been investigated. The first, which is analgous to the production and segregation of secretory proteins by the endoplasmic reticulum (Blobel 1976), involves the synthesis of polypeptides at the nuclear surface and their simulataneous translocation across the envelope. This hypothesis, however, is not consistent with the fact that extracted nuclear proteins are able to re-enter the nucleus following injection into the cytoplasm, as first shown by Bonner (1975b). In addition, Feldherr (1975) found that large endogenous polypeptides (mol. wt. > 94,000) continued to enter oocyte nuclei for several hours after protein synthesis had been inhibited with puromycin, demonstrating that synthesis and translocation are not coupled.

A second possibility is that precursor forms of nuclear proteins, which contain additional terminal polypeptides necessary for membrane transport, are initially sythesized in the cytoplasm. These molecules would then penetrate the envelope and, concurrently, be transformed into mature proteins by enzymatic cleavage of the terminal region. One example of such a mechanism is the uptake of F_1-ATPase subunits by the mitochondria (Maccecchini et al. 1979). The demonstraton by Bonner (1975b) that extracted nuclear polypeptides can re-enter the nuclei, argues against the existence of precursor molecules; however, because these were long-term experiments, the possibility exists that molecules which are normally transported across the envelope could have entered the nuclei by an alternative mechanism, such as diffusion. This criticism is not substantiated by the short-term injection studies, reported in this paper, which show that extracted and endogenous N1 and N2 distribute similarly between the nucleus and cytoplasm. Furthermore, nuclear and cytoplasmic N1 and N2 comigrate on two-dimensional SDS-polacrylamide gels, indicating that they have comparable charges and molecular weights. This is also inconsistent with the idea that there are cytoplasmic precursor polypeptides.

Thus, there is currently no evidence to support the concept that proteins can directly penetrate the membranes of the envelope, although mechanisms other than those considered could be functional.

SUMMARY

The major objective of the experiments reported above was to investigate possible mechanisms involved in regulating nucleoplasmic exchanges of endogenous proteins. Our initial studies centered around the function of the nuclear envelope. The approach used in these studies was to alter the physical properties of the envelope by puncturing nuclei in intact cells, and following any subsequent changes in protein distribution. Surprisingly, disruption of the envelope did not markedly affect the protein composition of the nuclei, indicating that the accumulation of specific polypeptides in this organelle is primarily due to selective binding. It was also determined that the envelope is not a major factor in regulating the rates at which many proteins accumulate in the nucleus. Considering that the envelope greatly restricts the movement of large exogenous molecules, it was especially surprising to find that the uptake rates of endogenous polypeptides with molecular weights of approximately 100,000 were unaffected by puncturing. The differences in the function of the envelope toward exogenous and endogenous molecules can be explained in either of two ways: first, that specific binding is rate-limiting for the nuclear uptake of endogenous molecules or, second, that endogenous proteins are, in some way, transported across the envelope. In this regard we investigated 1) the effect of nuclear binding on the uptake of specific polypeptides, and 2) the possibility that there is direct transport of proteins across the membranes of the envelope. The results showed that neither nuclear binding nor transport across the membranes function in regulating nucleocytoplasmic exchanges. Other mechanisms, including transport through the pores, are currently being considered.

ACKNOWLEDGEMENTS

This work was supported by grants GM 21531 from the National Institutes of Health and PCM 8003697 from the National Science Foundation.

Blobel G (1976). Synthesis and segregation of secretory proteins: The signal hypothesis. In Brinkley BR, Porter KR (eds): "International Cell Biology," New York: The Rockefeller University Press, p 318.
Bonner WM (1975a). Protein migration into nuclei. I. Frog oocyte nuclei in vivo accumulate microinjected histones, allow entry to small proteins, and exclude large proteins. J Cell Biol 64:421.
Bonner WM (1975b). Protein migration into nuclei. II. Frog oocyte nuclei accumulate a class of microinjected oocyte cytoplasmic proteins. J Cell Biol 64:431.
DeRobertis EM, Longthorne RF, Gurdon JB (1978). Intracellular migration of nuclear proteins in Xenopus oocytes. Nature 272:254.
Feldherr CM (1969). A comparative study of nucleocytoplasmic interactions. J Cell Biol 42:841.

Feldherr, CM (1975). The uptake of endogenous proteins by oocyte nuclei. Exp Cell Res 93:411.

Feldherr CM, Pomerantz J (1978). Mechanism for the selection of nuclear polypeptides in Xenopus oocytes. J Cell Biol 78:168.

Feldherr CM, Ogburn JA (1980). Mechanism for the selection of nuclear polypeptides in Xenopus oocytes. II. Two-dimensional gel analysis. J Cell Biol 87:589.

Kressmann A, Birnstiel ML (1980). Surrogate genetics in the frog oocyte. In Celis JE, Graessmann A, Loyter A (eds): "Transfer of Cell Constituents into Eukaryotic Cells," New York and London: Plenum Press, p 383.

Lusis AJ, Swank RT (1980). Regulation of location of intracellular proteins. In Prescott DM, Goldstein L (eds): "Cell Biology: A Comprehensive Treatise," New York: Academic Press, Vol. 4, p 339.

Maccecchini M-L, Rudin Y, Blobel G, Schatz G (1979). Import of proteins into mitochrondria: precursor forms of the extramitochondrially made F_1-ATPase subunits in yeast. Proc. Natl. Acad. Sci USA 76:343.

O'Farrell PH, O'Farrell PZ (1977). Two-dimensional polyacrylamide gel electrophoretic fractionation. In Stein G, Stein, J, Kleinsmith LJ (eds): "Methods in Cell Biology: Chromatin and Chromosomal Protein Research I," New York: Academic Press, Vol. 16, p 407.

Paine PL, Feldherr CM (1972). Nucleocytoplasmic exchanges of macromolecules. Exp Cell Res 74:81.

Paine PL, Moore C, Horowitz SB (1975). Nuclear envelope permeability. Nature 254:109.

MECHANISMS OF NUCLEAR PROTEIN CONCENTRATION

Philip L. Paine

Department of Biology
Michigan Cancer Foundation
Detroit, MI 48201

The eukaryotic cell exhibits a wide range of nucleocytoplasmic distributions of proteins. Although proteins are synthesized in the cytoplasm, some are found more or less equally in nucleus and cytoplasm, others are localized almost exclusively in the cytoplasm, and a third class - the so-called "N-proteins" - concentrate highly in the nucleus (Bonner 1975; DeRobertis et al. 1978; Paine, Horowitz 1980). We have undertaken a quantitative study of the N-proteins. Some N-proteins are undoubtedly structural members of the nuclear matrix, providing a framework or skeleton for maintenance of overall nuclear form and a scaffold for spatial and temporal coordination of intranuclear synthetic and transport processes. Others must diffuse within the aqueous interstices of the matrix, and a dynamic equilibrium between structural and diffusive forms may modulate the functional activities of specific N-proteins.

Nucleus/cytoplasm concentration ratios greater than 100 are maintained during interphase by some N-proteins. But at nuclear breakdown during cell division, N-proteins are released to the cytoplasm, where they distribute evenly; subsequently, they concentrate again in the reforming daughter nuclei (Krohne, Franke, 1980; Gerace, Blobel 1980). When isolated N-proteins are injected into the cytoplasm of interphase cells, they reaccumulate in the host nucleus at levels similar to their original distributions (Bonner 1975; Mills et al. 1980). These behaviors of N-proteins seem paradoxical since, by routine methods of cell fractionation, biochemistry, and histochemistry, they appear to be quite soluble (i.e., diffusible) to lack specific intranuclear sites of localization (DeRobertis et al. 1978; Mills et al. 1980; Krohne, Franke 1980) and to not be appreciably influenced by the nuclear envelope (Feldherr 1980).

Three classes of mechanism can, in principle, account for protein accumulation in the nucleus relative to the cytoplasm: (1) energy-dependent nuclear envelope transport, (2) adsorption or covalent linkage to nuclear structural elements, and (3) exclusion of diffusive proteins from water of the cytoplasm. Unfortunately, experimental data essential for distinguishing and measuring contributions of these mechanisms have not been available. This situation stems directly from the small size of cells and nuclei and the resulting technical limitations involved in quantitatively measuring the intracellular distributions of proteins. In addition, the physical states - structural (nondiffusive) or diffusive - of proteins within the living cell have been indeterminable. Both the intracellular distributions and the in vivo physical states of proteins are unavoidably altered by routine procedures for cell disruption and nuclear isolation.

We describe here an experimental system developed in order to alleviate these technical difficulties. It enables us to unambiguously measure the intracellular distribution, and the concentrations of bound and free forms of virtually any protein within a single living cell. (a) Microinjection of a gelatin reference phase establishes a defined intracellular compartment which equilibrates with diffusive proteins (Horowitz et al. 1979; Paine, Horowitz 1980); (b) cryomicrodissection at -45°C allows nucleus, reference phase, and cytoplasmic samples to be separated without artifactual solute redistribution or cross-contamination of contents (Century et al. 1970; Frank, Horowitz 1975); and (c) high-resolution two-dimensional microelectrophoresis makes possible the separation and quantitation of hundreds of individual polypeptides from a single isolated nucleus, reference phase, or cytoplasm sample.

Our experimental cell is the full grown oocyte of Xenopus laevis. The giant size of the oocyte (~1200 microns diameter) and its nucleus (~400 microns diameter) makes it ideal for reference phase microinjection and subsequent cryomicrodissection. Application of the intracellular reference phase and cryomicrodissection methods are illustrated in Figure 1. A 10-20% aqueous solution of purified gelatin is microinjected into the cell with a glass micropipette (Fig. 1a), and gelled in place by briefly cooling the cell to 13°C. The gelled gelatin, occupying 4-6% of the cell volume, is a fibrous protein network which excludes organelles and nondiffusive proteins, but has the solvent properties of ordinary water (Horowitz et al. 1979) and comes into equilibrium with proteins and other solutes which are diffusive within the living cell (Fig. 1b). After diffusional equilibrium is reached, the cell is mounted on a brass block and frozen to liquid nitrogen temperature (Fig. 1c). The cell is thenceforth maintained at temperatures not exceeding -45°C to prevent artifactual redistribution of proteins,

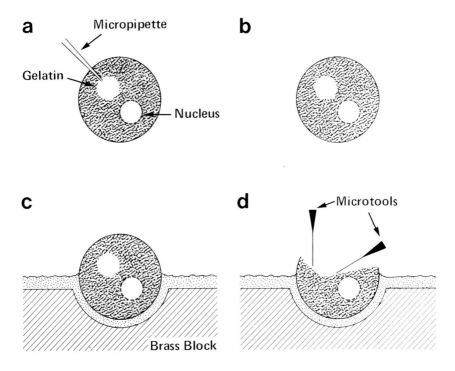

Fig. 1. Steps in the application of an intracellular gelatin reference phase and cryomicrodissection to the amphibian oocyte. See text for discussion.

Fig. 2 (opposite page). Two-dimensional polypeptide maps of the nucleus (a), reference phase (b), and cytoplasmic sample (c) from one Xenopus laevis oocyte. Molecular weight standards along the sides of gels (from top to bottom): BSA, 67,000; G6PD, 57,000; actin, 43,000; aldolase, 40,000; BCA, 30,000; myoglobin, 17,000. pH gradient along the tops of gels.

while the nucleus, reference phase, and cytoplasmic samples are manually isolated with fine microtools (Fig. 1d), cleaned, weighed, and stored in tiny aluminum foil packets. Individual cryomicrodissected samples of known water content are dissolved in lysis buffer, loaded onto tube gels (4% acrylamide in 50 μl micropipettes), and subjected to isoelectric focussing. Subsequently, the tube gels are loaded onto thin SDS slab gels, electrophoresed in the second dimension, and the resultant two-dimensional polypeptide maps visualized by silver staining.

Figure 2a-c shows 2-D maps of a nucleus, reference phase, and cytoplasmic sample (top to bottom) from one oocyte. Maps of nucleus and cytoplasm share some proteins but are very different. The pattern of proteins in the reference phase - those which are diffusive within the cell - more closely resembles the nucleus than cytoplasm. However, upon close examination, important differences between nucleus and reference phase are also seen; prominent nuclear proteins are totally absent from the reference phase.

For quantitation of individual polypeptides, an automated gel analysis system (Merril et al. 1981) was used. The system utilizes a high-speed scanning densitometer which records a complete two-dimensional optical density scan of each gel into disc storage. Gel images can be recalled, displayed, and manipulated on a television screen, and the integrated density of any individual polypeptide spot, with background corrected, determined using an image processor and a computer. Background-corrected density values were divided by the known water contents of cryomicrodissected samples to give measures of the polypeptide concentrations in the nucleus, reference phase, or cytoplasm. From the gels of Figure 2, 90 individual polypeptide spots [positions indicated by numbers superimposed on the nuclear pattern (Fig. 3)] were selected for quantitative analysis.

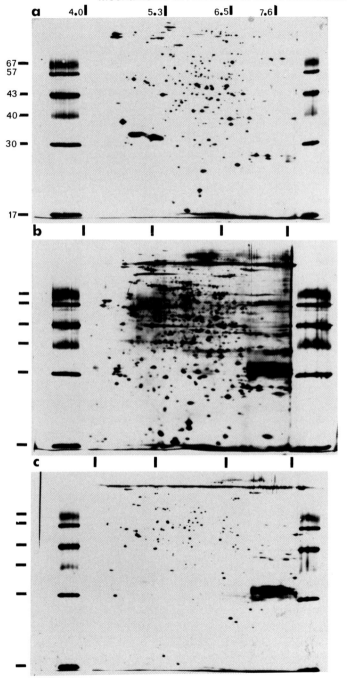

From the 90 proteins measured, we identify 35 with nuclear/cytoplasmic concentration ratios (N/C) greater than 10 (Table 1). This class appears heterogeneous on the denaturing gel system, displaying N/C values from 10 to more than 600, and sizes ranging from 20,000 to 150,000 daltons. The isoelectric points of those polypeptides we find to be most concentrated in the nucleus [i.e., 60-61 (nucleoplasmin) (Mills et al. 1980), 62, 74 and 75] are 5.3 or less, implying a role for protein negative charge in nuclear concentration.

TABLE I

Polypeptides ("N-Proteins") with N/C>10:

Polypeptide	N/C [1]	RP/C [1]	Polypeptide	N/C	RP/C
(N-Proteins detected in the RP)					
4	11.3	2.5	57	> 20	2.6
13	23.0	0.3	58	> 11	1.1
49	> 125	> 10	59	> 15	1.9
50	> 15	> 3.2	65	> 20	> 1.6
51	> 30	> 9.2	66	> 40	> 3.8
52	12.3	3.7	74	> 250	> 0.2
53	> 60	> 12.7	75	> 190	> 0.9
56	> 20	> 5.2			
(N-Proteins not detected in the RP)					
60	> 600	–	77	> 20	–
61	> 590	–	78	> 75	–
62	> 210	–	79	> 20	–
63	> 50	–	80	> 40	–
64	> 40	–	81	> 50	–
67	> 25	–	82	> 15	–
68	> 20	–	83	> 100	–
70	> 20	–	84	> 50	–
71	> 100	–	85	> 135	–
73	> 20	–	86	> 30	–

The majority of the N-proteins we have measured are not detected in the reference phase, indicating very low concentrations of diffusive forms of these proteins. It is unlikely that the nuclear envelope, with its known permeability to proteins (Paine et al. 1975), can actively maintain cytoplasmic diffusive protein concentrations at levels this low, and we believe that the nuclear concentration of those N-proteins not found in the reference phase is primarily due to nuclear binding or incorporation into nuclear structural elements. [In this regard, Feldherr (1980) has reported

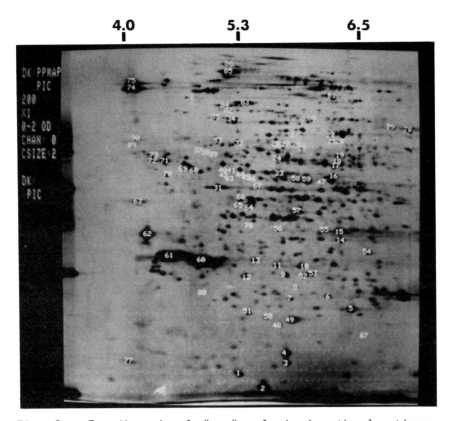

Fig. 3. Two-dimensional "map" gel showing the locations of polypeptides selected for quantitative computer analysis. This is the television image of the nuclear gel in Fig. 2a, recalled from disc storage. The amount of each indicated polypeptide was determined on each 2-D gel (nucleus, reference phase, and cytoplasm); division by the known water contents of the samples loaded onto the gels yielded the in vivo concentrations in each cellular compartment. Those proteins whose nucleocytoplasmic concentration ratios exceeded 10 were arbitrarily classed as N-proteins and their data are presented in Table 1.

of some N-proteins are not detectably changed after increasing nuclear envelope permeability by repeated puncturing of the envelope with glass needles.]

Other N-proteins are detectable in the reference phase, and the distributions of a few of these implicate exclusion from cytoplasmic water as another significant mechanism of nuclear accumulation. Inert exogenous macromolecules, such as myoglobin and neutral dextrans (Paine et al. 1975), are known to equilibrate diffusively after microinjection into oocytes with N/C ~ reference phase/C ~ 2-3, reflecting reduced solubility in, or exclusion from, 50-65% of the water of cytoplasm relative to the water of the nucleus or reference phase. For some N-proteins (polypeptides 49, 51, 53, and 56), the reference phase/C values exceed 2-3, implying that cytoplasmic exclusion may make even larger contributions to the nuclear accumulation of these specific cellular proteins.

ACKNOWLEDGEMENTS

We thank Drs. D. Goldman and C.R. Merril for their assistance and the use of their facilities at NIMH for the quantitative computer analysis of our gels. This work was supported by NIH grants GM 26734 and GM 19548, and by an institutional grant from The United Foundation of Greater Detroit.

Bonner WM (1975). Protein migration into nuclei. II. Frog oocyte nuclei accumulate a class of microinjected oocyte nuclear proteins and exclude a class of microinjected oocyte cytoplasmic proteins. J Cell Biol 64:431.
Century TJ, Fenichel IR, Horowitz SB (1970). The concentrations of water, sodium, and potassium in the nucleus and cytoplasm of amphibian oocytes. J Cell Sci 7:5.
DeRobertis EM, Longthorne RF, Gurdon JB (1978). Intracellular migration of nuclear proteins in Xenopus oocytes. Nature (London) 272:254.
Feldherr CM (1981). Mechanism for the selection of nuclear polypeptides in Xenopus oocytes. II. Two-dimensional gel analysis. J Cell Biol 87:589.
Frank M, Horowitz SB (1975). Nucleocytoplasmic transport and distribution of an amino acid in situ. J Cell Sci 19:127.
Gerace L, Blobel G (1980). The nuclear envelope lamina is reversibly depolymerized during mitosis. Cell 19:277.
Horowitz SB, Paine PL, Tluczek L, Reynhout JK (1979). Reference phase analysis of free and bound intracellular

solutes. I. Sodium and potassium in amphibian oocytes. Biophys J 25:33.

Krohne G, Franke WW (1980). A major soluble acidic protein located in nuclei of diverse vertebrate species. Exp Cell Res 129:167.

Merril CR, Goldman D, Sedman SA, Ebert MH (1980). Ultrasensitive strain for proteins in polyacrylamide gels shows regional variation in cerebrospinal fluid proteins. Science 211:1437.

Mills AD, Laskey RA, Black P, DeRobertis EM (1980). An acidic protein which assembles nucleosomes in vitro is the most abundant protein in Xenopus oocyte nuclei. J Mol Biol 139:561.

Paine PL, Moore LC, Horowitz SB (1975). Nuclear envelope permeability. Nature 254:109.

Paine PL, Horowitz SB (1980). The movement of material between nucleus and cytoplasm. In Prescott DM, Goldstein L (eds): Cell Biology: A Comprehensive Treatise, Vol. IV, "Gene Expression: Translation and the Behavior of Proteins", New York: Academic Press, p 299.

RNA TRANSPORT STUDIES IN A CELL-FREE SYSTEM

Dorothy E. Schumm

Department of Physiological Chemistry
The Ohio State University College of Medicine
Columbus, Ohio 43210

The significance of post-transcriptional events in the modulation of genetic expression has been recognized for some time. There have been many studies on capping, polyadenylation and messenger RNA splicing. However, until recently, the role of the nuclear envelope in this process has gone unrecognized and unstudied. The existence of this conference attests to the growing realization of its importance.

Before RNA can function as a messenger in protein synthesis, it must be properly processed and then transported from the nucleus. If other processing events are inhibited, nucleocytoplasmic transport of messenger RNA does not occur. For example, disruption of helical structures by proflavine or inhibition of polyadenylation by cordycepin both cause decreases in RNA transport. The restriction of some RNAs to the nucleus is not merely a matter of size since messengers as large as 75S are transported in the silkworm and the procursor for α-globulin, which is only 800 nucleotides long, is retained. The phenomenon of nuclear restriction is one of the events which is altered early in chemical carcinogenesis (Patel et al. 1979; Schumm et al. 1977). RNA transport is also affected by cytoplasmic proteins. Some of these enter the nucleus and serve as enzymes for the processing events, some are bound to the RNA when it emerges from the nucleus, while others modulate transport by a still unknown mechanism. It is difficult to study the proteins connected with transport in intact cells. Therefore during the past 10 years, Dr. Thomas Webb and I have worked on the development of a cell-free system in which to study the nucleocytoplasmic transport of both messenger and ribosomal RNA. The cell-free system consists of 30-min prelabeled rat liver nuclei incubated in a surrogate cytoplasm, containing 12 mg/ml of dialyzed 105,000 x g supernatant, 50 mM Tris buffer (pH 7.5), 250 mM sucrose, 25 mM KCl, 2.5 mM $MgCl_2$, 0.5 mM $CaCl_2$, 0.3 mM $MnCl_2$, 2.5 mM Na_2HPO_4, 5 mM NaCl, 5.0 mM spermidine, 2.5 mM dithiothreitol, 2.0 mM ATP, 2.5 mM phosphoenol pyruvate, 6.4 units/ml pyruvate kinase 0.25 mM GTP, 0.1 mM methionine and 500 µg/ml of low molecular weight yeast RNA.

Both in vivo and in vitro, RNA is released as a ribonucleoprotein particle with a density of 1.47 g/cc. The transported RNA is heterodisperse, centrifuging in a range from 8 to 18S on sucrose density gradients. A similar size distribution is seen in vivo. After 30 min of incubation at 30°C, 5 to 6% of the nuclear counts are released as acid precipitable RNA, 65% of which contain poly(A) tracts. We have recently shown that the transported RNA is capable of forming initiation complexes with ribosomes in a cell-free translation system. We have also shown that one of the messengers which is released is that for rat albumin (Palayoor et al. 1981). The amount released is similar to that observed in vivo.

Table 1

Source of mRNA	Leucine incorporation	Ratio albumin/total
Polysomes	13908	.32
Released in vivo	13627	.34

Since RNA can be released from detergent washed nuclei, it is apparent that the messenger RNA observed is being transported from the inside of the nucleus and is not merely released from the outside of the nuclear envelope. Labeling studies support this conclusion.

The transport of RNA from the nucleus into the supernatant of the cell-free system is dependent on the presence of a hydrolyzable nucleotide triphosphate. The exact function of the energy is unclear at this time but is likely to include multiple sites involving both processing and RNA translocation. Hydrolysis of the β-γ bond of ATP is necessary. Nonhydrolyzable ATP analogs and other high energy compounds such as pyrophosphate and cyclic nucleotides cannot serve as energy sources. Since this is the case, it is unlikely that simple chelation of metal ions is taking place. The energy is not being used for continuing RNA synthesis since large amounts of actinomycin D do not inhibit transport in this system, nor is it being used for polyadenylation. Smuckler's group (Clawson et al. 1978) has shown that inhibition of nucleotide hydrolysis by copper sulfate, zinc chloride, silver nitrate and rotenone also inhibited RNA transport. We have shown (Schumm, Webb 1975) a similar effect for beryllium and colchicine.

Recently we have concentrated most of our efforts on isolation and purification of the cytoplasmic proteins which modulate RNA transport in the cell-free system. We have found that the transport protein(s) located in the cytosol recycles in and out of the nucleus, but at any one time only 1% of the factor can be isolated from the nuclear fraction (Moffett, Webb 1981). The majority of the factors are found in the polysomes. They are only loosely bound and can be removed by a 0.5 M KCl wash but not by 0.25 M KCl. Sephadex chromatography indicates a molecular weight of around 40,000. The factors are capable of binding to an affinity column containing single-stranded DNA but have a higher affinity for messenger RNA. They have been purified more than 1000-fold but several bands are still observed on acrylamide gels. The difficulty in achieving purification is not surprising since these proteins constitute only 1% of the total cytoplasmic proteins. They are precipitated by ammonium sulfate and surprisingly by streptomycin sulfate. They are inactivated by temperatures greater than 50°C. Inhibition of protein synthesis indicates that they turn over very rapidly. Treatment for 5 hr with pactomycin or cycloheximide causes only a slight reduction of the total cytosol protein but causes a 33-50% reduction in the transport factors (Yannarell et al. 1976).

The mechanism of action of these proteins is unclear. The molecular weight excludes ribonuclease. However, studies with cyclic nucleotides suggest that a protein kinase may be involved (Schumm, Webb 1978). The addition of physiological concentrations of cAMP causes a 50% increase in the amount of acid precipitable RNA released from 30-min prelabeled nuclei in a cell-free system derived from resting rat liver.

Table 2

Amount cAMP added (M)	% Nuclear RNA released		
	Resting liver		cAMP treated
	12 mg/ml	2 mg/ml	
None	4.3	1.2	3.0
10^{-4}	4.6	1.2	3.0
10^{-5}	5.5	1.3	3.1
10^{-6}	7.7	1.2	3.1
10^{-7}	5.6	1.1	3.0
10^{-8}	4.8	1.2	3.1
10^{-9}	4.5	1.2	3.1
Theophylline 10^{-2} M	4.3		3.2
10^{-6} plus 10^{-2} M Theophylline	7.7		3.1
10^{-6} minus ATP, PEP, GTP	0.3		0.2

The effect of cAMP is achieved by doubling the effective concentration of the mRNA transport factor(s). This presumably occurs via phosphorylation by a cAMP-dependent protein kinase. The cAMP effect is removed if cytosol proteins are precipitated by streptomycin, if the cytosol concentration is lowered to 2 mg/ml or if the intact animal is treated for 30 min with 50 mg/kg of cAMP prior to preparation of the cytosol. The latter effect is attributed to the in vivo translocation of the cAMP binding protein to the nucleus. Thus, it would be absent in such cytosol and be unavailable for binding cAMP and mediating its effects. Similar results have been obtained for cGMP where the optimal concentration was found to be 10^{-8} M. Incidentally, cyclic nucleotides have no effect on ribosomal RNA transport.

There are two other agents which we have studied which modify the nuclear envelope and which also modify RNA transport. One of these is temperature. Nagel and Wunderlich (1977), using Tetrahymena have found lipid clustering and decreased membrane fluidity when the organism was grown at reduced temperatures. This correlated with reduced RNA transport at these temperatures. We have found a similar temperature discontinuity in the cell-free system. At temperatures below 18°C, there was little RNA transport. Above this temperature, transport increased linearly with increasing temperature.

Another modifier of both the nuclear envelope and RNA transport is insulin. Recently, insulin has been shown to be internalized and bound to high affinity sites on the nuclear membrane, the number of which appear to be regulated by the plasma insulin levels in normal animals (Goidl 1979). The physiological function of this binding is unknown. Using our cell-free system, we investigated the effect of insulin on RNA transport. Depending on the concentration, insulin either stimulates or inhibits RNA transport. At 3×10^{-7} M insulin, RNA transport was increased by approximately 50%. At higher concentrations, insulin depressed transport. This effect occurred with nuclei isolated from normal rat liver. However, in nuclei from the genetically obese insulin-resistant Zucker rat, there was no increase in RNA transport at any insulin concentration tested. The lean heterozygote showed an intermediate response. Measurement of the poly(A) content of the released RNA showed an increase from 42% in the control to over 60% in systems to which an optimal concentration of insulin had been added. This suggests that essentially all of the additional RNA transported in the presence of insulin consists of poly(A) containing RNA. As with the cyclic nucleotides, insulin showed no effect on ribosomal RNA transport.

Under normal conditions, RNA transport rather than RNA processing appears to be the rate limiting step. Heavy RNA in the nucleus is converted to lighter species which accumulate and only slowly are released to the cytosol. In the presence of insulin, there is a significant reduction in the retention of the lighter RNAs. Thus, it appears that insulin exerts its primary effect on RNA transport and not on processing. In agreement with this is the observation that treatment of nuclei with Triton X-100 to remove the outer nuclear membrane completely abolishes the insulin response.

The known enhancing effect of insulin on the hepatic content of mRNAs coding for albumin, α_2-globulin, tryosine amino transferase and certain ribosomal proteins suggests that insulin may be acting, not as a general stimulator of RNA transport, but rather as a stimulator of the transport of specific mRNA species. Such an effect could be mediated if processed gene products were, for example, directed to specific nuclear pores and insulin acted on these pores either as a stimulator or inhibitor. The possibility that the nuclear envelope is one site of action for other hormones deserves further consideration.

Clawson GA, Koplitz M, Castler-Schechter B, Smuckler EA (1978). Energy utilization and RNA transport: their interdependence. Biochem 17:3747.
Goidl JA (1979). Insulin binding to isolated liver nuclei from obese and lean mice. Biochemistry 18:3674.
Nagel WC, Wunderlich F (1977). Effect of temperature on nuclear membranes and nucleocytoplasmic RNA transport in Tetrahymena grown at different temperatures. J Membrane Biol 32:151.
Moffett RB, Webb TE (1981). Regulated transport of messenger ribonucleic acid from isolated liver nuclei by nucleic acid binding proteins. Biochemistry 20:3253.
Palayoor T, Schumm DE, Webb TE (1981). Transport of functional messenger RNA from liver nuclei in a reconstituted cell-free system. Biochim Biophys Acta 654:201.
Patel NT, Folse DA, Holoubek V (1979). Release of repetitive nuclear RNA into the cytoplasm in liver of rats fed 3'-methyl-4-methylaminoazobenzene. Cancer Res 39:4460.
Schumm DE, Hanausek-Walaszek M, Yannarell A, Webb TE (1977). Changes in nuclear RNA transport incident to carcinogenesis. Eur J Cancer 13:139.
Schumm DE, Webb TE (1975). Differential effect of ATP on RNA release from nuclei of normal and neoplastic liver. Biochem Biophys Res Commun 67:706.
Schumm DE, Webb TE (1978). Effect of adenosine 3':5-monophosphate and guanosine 3':5'-monophosphate on RNA release from isolated nuclei. J Biol Chem 253:8513.
Schumm DE, Webb TE (1981). Insulin-modulated transport of RNA from isolated liver nuclei. Arch Biochem 210:275.
Yannarell A, Schumm DE, Webb TE (1976). Nature of the facilitated messenger ribonucleic acid transport from isolated nuclei. Biochem J 154:379.

COMPARISON OF METHODS FOR STUDYING RNA EFFLUX FROM ISOLATED NUCLEI

Paul S. Agutter, Ph.D.

Department of Biological Sciences
Napier College
Colinton Rd.
Edinburgh EH10 5DT, U. K.

The main aim of our work for the past seven years has been to elucidate the mechanism of nucleocytoplasmic RNA transport, with particular reference to mammalian liver. This aim is justified by indications from a number of sources that this mechanism is biochemically non-trivial (Paine et al. 1975; Lichtenstein, Shapot 1976); by the suggestion that in at least some eukaryotic systems, protein biosynthesis might be regulated at the RNA transport stage; and by clear demonstrations in several laboratories that the mechanism is markedly different in hepatoma cells from that in hepatocytes (Schumm et al. 1973a; Schumm et al. 1977; Shearer, Smuckler 1972). Most of the advances to date in the elucidation of the process in mammalian liver cells will already be familiar. These include demonstrations that ATP hydrolysis supplies the energy for transport (Agutter et al. 1976, 1979; Clawson et al. 1980a, b; Murty et al. 1980); that the transported RNA must have an organized tertiary structure, and that the binding and release of this RNA to and from the transport apparatus is associated respectively with phosphate release and with ATP binding (Agutter, Ramsay 1979; McDonald, Agutter 1980); and that there are protein factors (as yet not fully characterized) in the cytoplasm that stimulate transport and possibly others that inhibit it (Yu et al. 1972; Schumm et al. 1973b; Racevskis, Webb 1974; Lemaire et al. 1981). Furthermore, the physical state of the lipid of the nuclear envelope seems not to be important in the regulation of transport in mammalian liver (Stuart et al. 1977), though in some lower eukaryotes, e.g., <u>Tetrahymena</u>, it does appear to have a role (Nagel, Wunderlich

1977). Currently, the possibility that the dynamics of the nuclear matrix are important in regulating RNA transport is receiving attention in our laboratory and in others (Herlan et al. 1979).

In the present paper, I intend to discuss experimental approaches to the study of RNA transport. If the mechanism is to be elucidated, then for most experiments the system studied must be the simplest adequate one. By "simplest," I mean not only the one of least complex composition, but also the one which involves the minimum number of uncontrolled factors and therefore enables investigators to modify conditions in quantifiable ways and to interpret results as unambiguously as possible. By "adequate" I mean one in which nuclear properties such as stability, permeability and RNA restriction correspond closely (that is, closely enough to give results that are not seriously disturbed by artifacts) to those in vivo. For RNA transport studies, this usually entails working with isolated nuclei in the simplest adequate medium, though in some cases compositionally simpler preparations can be used, e.g., RNA-rich nuclear matrices prepared by methods such as those of Pogo and his co-workers (Long et al. 1979). For studies of the transport system itself, nuclear envelope and matrix preparations often seem to be appropriate. I shall begin with a brief discussion of envelope and matrix preparations, and then devote most of this paper to a comparative study of incubation media that have been used for experiments on RNA efflux from whole isolated nuclei.

ENVELOPE AND MATRIX PREPARATIONS

Because of early evidence that ATP had a role in RNA transport (Ishikawa et al. 1969) the hypothesis was advanced that the process involved ATP hydrolysis at or near the surface of the nucleus, supplying energy for transport. To test this hypothesis, it was necessary to characterize one or more nucleoside triphosphatase (NTPase) activities in the nuclear envelope, and this required nuclear envelopes that were not only morphologically intact but also had high NTPase activity. It was found that such activity was largely destroyed when the high pH required in isolation procedure of Kay et al. (1972) was used; also exposure to high ionic strength media as used in the methods of Franke et al. (1970) and Monneron (1974) inactivated the ATPase more or less rapidly. The development of a modification of the procedure of Kay et al. (1972) by Harris and Milne (1974) yielded nuclear envelopes in which the activity was high, though somewhat variable, and this resolved the problem. Using this method, we were able to characterize the NTPase satisfactorily.

There is now widespread interest in the role of the matrix in several aspects of nuclear function; some of these are discussed in other paper in this Symposium.

A specific role for the matrix in RNA transport was postulated by Wunderlich and his coworkers on the basis of their experiments with Tetrahymena (Herlan et al. 1979). To determine whether the matrix has such a role in the mammalian liver (Agutter, Richardson 1980), two experimental criteria must be met: the matrices isolated must be largely intact and disaggregated, and they must contain most or all of the nuclear RNA. The first criterion is established by the conclusion of Wunderlich's group that matrix expansion and contraction control RNA release; therefore, expansion and contraction must be studied, and in the interests of statistical reliability this requires a large population of intact, separate individuals. Berezney (1979) has drawn attention to the difficulty of meeting this criterion, but in our experience, the method of Pogo and his co-workers (Long et al. 1979) goes a considerable way towards it. The second criterion can be met by using serine protease inhibitors in the isolation media (Miller et al. 1978; Berezney, Buchholtz 1978; Agutter, Birchall 1979), but these reagents inhibit the NTPase; this places obvious limitations on the utility of methods using reagents of this kind in RNA transport studies.

STUDIES ON WHOLE ISOLATED NUCLEI

I have emphasized the importance of the choice of isolation procedure for the study of envelopes and of matrices with respect to RNA transport; the same importance is obviously attached to the isolation of the nuclei themselves. The method of Blobel and Potter (1966) is the most suitable but gentle homogenization conditions and crucial in order to minimize damage to the nuclei.

Given the appropriate isolation procedure, it remains to develop the simplest adequate incubation medium. In our laboratory, several of the media that have been described in the literature have been compared in terms of the following criteria of adequacy:

a) characterization, e.g., using density gradient centrifugation, of the RNA released in the presence and absence of ATP;

b) determination of the amount of DNA released;

c) characterization, by sodium dodecyl sulfate-polyacrylamide gel electrophoresis (SDS-PAGE), of the proteins released;

d) estimation, by phase-contrast microscopy, of the extent to which normal nuclear morphology is retained during the incubations. The indicators of morphological change considered were nuclear swelling and shrinkage, aggregation, lysis, granulation of the nuclear contents, and nucleolar disruption.

Although a large number of different media was tested, I shall confine my account to a comparison of three: media A (Ishikawa et al. 1969), B (Chatterjee, Weissbach 1973), and C (Yu et al. 1972). In the case of this last medium, I shall also mention two modifications: C_1 (Ishikawa et al. 1978) and C_2 (Agutter et al. 1976, 1979). The compositions are summarized in Table 1. Other media have characteristics predictable from the results given in this study.

Characterization of RNA Released

Table 2 shows the percentages of total nuclear RNA released in the presence and absence of ATP during 5 min and 20 min incubations. The results are based not simply on the cpm released, but also on the cpm/A_{260} ratios of the material eluted into the medium and of total nuclear RNA. In general, because the rats were injected with [^{3}H]uridine (1 μC_1/g body weight) 45 min before the removal of the liver and therefore the label was largely confined to mRNA and its precursors (Schumm, Webb 1978), the nuclei contained a large pool of unlabeled RNA (mainly ribosomal and pre-ribosomal).

Table 1

Composition of Media Compared

Medium	Tris-HCl (mM)	Sucrose (mM)	Thiol reagent (mM)	Spermidine (mM)	KCl (mM)	MgCl$_2$ (mM)	CaCl$_2$ (mM)	MnCl$_2$ (mM)	Yeast RNA	Cytoplasmic protein (mg/ml)	ATP regenerating system (mg/ml)
A	10	880	-	-	-	5.0	-	-	-	-	-
B	50	250	2.0	-	25	4.0	-	-	-	-	-
C	50	-	2.0	5.0	25	2.5	0.5	0.3	0.3	15	+
C$_1$	50	-	2.0	5.0	25	5.0	-	-	0.025	5	+
C$_2$	10	-	5.0	5.0	-	2.5	0.5	0.3	0.3	-	-

The references for the media are given in the text. All media were adjusted to pH 7.5 (35°C). The ATP-regenerating system comprised 2.5 mM P$_i$; the ATP-regenerating system comprised 2.5 mM phosphoenolpyrurate and 35 units pyrurate kinase. The thiol reagent was dithioerythritol in media B, C and C$_1$, and 2-mercaptoethanol in medium C$_2$.

Table 2

RNA Release into Various Incubation Media

(a) Medium		5 min incubation		20 min incubation	
		-ATP	+ATP	-ATP	+ATP
	A	2.7 + 0.4	12 + 2	9.2 + 1.4	25 + 5
	B	9.3 + 1.9	17 + 3	16 + 4	36 + 5
	C	Trace	5.2 + 1.1	2.1 + 0.3	11 + 2
	C_1	Trace	6.4 + 2.1	2.8 + 1.0	15 + 3
	C_2	Trace	1.7 + 0.3	0.9 + 0.5	3.2 + 0.9
(b)	A	ND	3.7 + 0	1.9 + 0.2	4.1 + 0.3
	B	2.6 + 0.2	2.8 + 0.5	2.1 + 0.1	3.2 + 0.5
	C	ND	5.3 + 0.6	ND	5.5 + 0.1
	C_1	ND	4.9 + 0.5	ND	4.6 + 0.4
	C_2	ND	5.8 + 0.3	ND	6.4 + 0.2

Radioactive counting of the RNA was performed as described by Agutter et al. (1979). Section (a) shows the percentage of total nuclear cpm released and (b) gives the cpm/A_{260} ratios (all x 10^{-5}). Results are means + SEM of the three triplicate determinations. Nuclear labeling was 12,340 + 2,163 cpm/10^7 nuclei, and total nuclear RNA contained 1.4 x 10^5 cpm/A_{260} unit. ND = not determined.

If conditions are sought for the study of mRNA release and a 45-min prelabel is used, then a high cpm/A_{260} ratio in the eluted material is one reasonable indication of success.

Taking this criterion alone, media C, C_1 and C_2 appear to maintain the RNA restriction of the nuclei for longer than do media A and B; also, the effects of ATP are more clearly marked in medium C and its variants. When the RNA released is further characterized by sucrose gradients (Fig. 1), it is clear that, in media and C_2, (i) most of the RNA released in the absence of ATP is around 4S, and the relatively high percentage of pseudouracil and 5-methylcytosine in this material suggests that it contains tRNA; (ii) in the presence of ATP, material of 8-18S (peaking around 10-12S)

Fig. 1. 5.0 ml sucrose density gradients of the RNA released from (a) media A and B in the presence of ATP, (b) media C and C_2 in the presence and absence of ATP, were developed for 16 hr at 100,000 x g and 20 x 0.25 ml fractions were counted. Ribosomal 28S and 18S RNA were run as standards. Results are means ± SEM of three triplicate determinations.

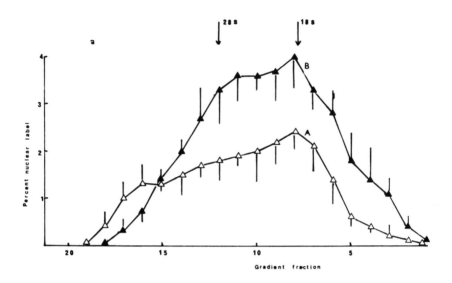

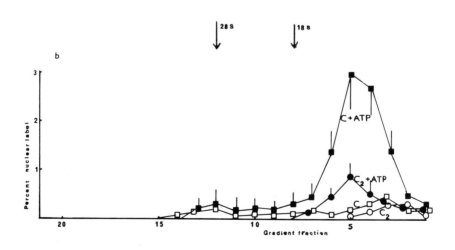

is released, and the high content (60-70%) of adenylated material in this, measured by binding to oligo-(dT)-Sepharose, suggests that it contains much mRNA or messenger precursors; (iii) in the presence of dialyzed cytosol plus an ATP-generating system - the essential difference between media C and C_2 - the extent of release of this 10-12S material is considerably enhanced. In media A and B, the RNA released in the presence of ATP showed a wider range of S values and the content of adenylated material was very variable, averaging about 40%.

These results suggest that medium C and its variants support nuclear restriction more satisfactorily and are therefore generally more suitable for the study of mRNA release than media A and B. However, two important problems remain. First, more needs to be understood about the differences between media and C_2 - i.e., about the effects of cytoplasmic protein and of the ATP-regenerating system. Second, some attempt must be made to determine the extent to which the RNA released resembles cytoplasmic messenger rather than messenger precursors.

The results shown in Table 3 pertain to the first problem. Cytoplasmic protein in the absence of an ATP-regenerating system has little effect on RNA release or on nuclear stability as indicated by DNA release. This is probably because this protein hydrolyzes ATP rather rapidly (50-60 μmol ATP/hr/mg protein) and therefore the ATP, unless regenerated, is rapidly hydrolyzed, cancelling out the stimulation of transport by the ATP itself. However, the regenerating system (specifically the phosphate) destabilizes nuclei markedly, in agreement with the findings of Schumm et al. (1973b), and the cytosol has the additional effect of protecting the nuclei against this destabilization. Preliminary results indicate that the protein fraction that protects against destabilization is excluded on G75 Sephadex, while the fraction responsible for the enhancement of the ATP-dependent release of RNA is retarded. The general conclusion from these findings is as follows: cytosol promotes RNA release; but it does so at the expense of complicating the conditions. Apart from the uncertain effects of high concentrations of dissolved protein on the activities of the ions in the medium, it has at least two apparently direct effects (protection against lysis and the promotion of RNA transport) on the nuclei, and necessitates the inclusion of the ATP regenerating system, which compli-

Table 3

Comparison of Variants of Medium C

2.0 ATP	Modifications regenerating system	Cytoplasmic protein	RNA released	DNA released
-	-	-	0.9 ± 0.5	0.4 ± 0.2
+	-	-	3.3 ± 0.2	0.9 ± 0.3
+	+	-	28 ± 4	15 ± 2
-	-	+	2.1 ± 0.3	1.6 ± 0.5
+	-	+	2.9 ± 0.6	2.1 ± 0.5
+	+	+	12 ± 2	2.4 ± 0.3

Percentages of total nuclear RNA and DNA released into variants of medium C during 20 min incubations at 37°C. Results are means ± SEM of three triplicate determinations.

cates the mixtures further. For this reason, we have been inclined to use the simpler medium C_2.

Table 4 pertains to the second problem, and the results which it summarizes support the conclusions drawn from Table 2 and Figure 1, and are in broad agreement with the findings of Schumm and Webb (1974) and of Clawson and Smuckler (1980). The released RNA was hybridized with total nuclear DNA in the presence and absence of total cytoplasmic RNA. On this basis, some 80-90% of the RNA released into medium C_2 in the presence of ATP resembles cytoplasmic material, compared to about 70% in media C and C_1 and lower percentages in media A and B. Results of this kind confirm our decision to use medium C_2 in our studies, but the well-known problems of interpreting gross hybridization data of this kind indicate that the evidence is not wholly compelling, nor is it likely to become so until such studies are repeated on cell types in which a small number of well-characterized cytoplasmic messenger types predominates.

Table 4

Hybridization Studies on Released RNA

Labeled RNA released in medium	cpm RNA bound to filter in absence of cytoplasmic RNA	In presence of cytoplasmic RNA	% of label competing with cytoplasmic RNA
A	2747 + 510	1428 + 372	48 + 13
B	3789 + 753	2463 + 307	35 + 8
C	2393 + 487	646 + 136	74 + 6
C_1	3323 + 665	1030 + 212	69 + 6
C_2	709 + 53	85 + 18	88 + 3
Total nuclear	12371 + 222	10420 + 204	16 + 2

Trichloracetic acid precipitates of the labeled RNA, released into the media during 5 min (A and B) or 20 min (C, C_1 and C_2) incubations in the presence of ATP, were dissolved in 0.1% (w/v) SDS, 15 mM sodium citrate, 150 mM NaCl, pH 7.5, and incubated for 40 + 2 hr with filter-bound total nuclear DNA in the presence or absence of a 50-fold excess of unlabeled total cytoplasmic RNA. During hybridization, the filters contained 470 + 40 µg and the labeled RNA did not exceed 3 ng; of this, 14-20% became bound in all cases, and sucrose gradients indicated that this was representative of the total labeled material. After hybridization, the bound label was removed with 3 x 3 mls 0.5% (w/v) SDS, 50 mM sodium citrate, 300 mM NaCl, and radioactivity counted. Results are means + SEM of three determinations.

Determination of DNA Released

When DNA leakage (determined by the method of Burton 1956) is used as an indicator of nuclear stability, the results (Table 5) indicate overall that conditions that maintain more normal nuclear RNA restriction also maintain stability. ATP has less effect on DNA leakage than it has on RNA release. The DNA released in media A and B is of low molecular weight, and agarose gels indicate a "chromatosome"

Table 5

DNA Release into the Media Used

Medium	Modification	Incubation time (min)	DNA released
A	-	5	6.6 ± 1.3
A	-	20	21 ± 4
B	-	5	16 ± 2
B	-	20	32 ± 5
C	-	20	2.4 ± 0.3
C_1	-	20	4.6 ± 0.5
C_2	-	20	0.9 ± 0.3
C_2	Spermidine omitted	20	18 ± 4
C_2	5 mMP_i added	20	13 ± 2
C_2	5 mMP_i + 15 mg/ml cytoplasmic protein added	20	1.6 ± 0.2
C_2	Ca^{2+} and Mn^{2+} omitted	20	2.9 ± 0.6
C_2	Ca^{2+} and Mn^{2+} omitted; Mg^{2+} increased to 5.0 mM	20	5.8 ± 0.9

Percentages of total nuclear DNA released into the media during incubation at 37°C in the presence of ATP. Results are means ± SEM of three triplicate determinations.

size (165-180 base pairs). In medium C and its variants, very little DNA is released, so the indication that it is again predominantly in this size-class must be regarded as provisional. DNA release probably results from the action of endogenous endonucleases in the preparation.

The reasons for the differences between the media with respect to nuclear stability deserve investigation. In Table 5, the effects of varying the ion contents of medium C_2 are summarized. Clearly, as other authors have indicated (Schumm, Webb 1974; Sato et al. 1977), spermidine is crucially important. Moreover, inorganic phosphate destabilizes nuclei in the absence of cytosol, and this is consistent with the effect of the ATP-regenerating system on RNA release (Table 3). The stabilizing effects of Ca^{2+} and Mn^{2+} have been described previously; here, again, the effects on DNA release are paralleled by effects on RNA release (Agutter 1980). The most surprising effect is that of Mg^{2+}. Because this ion activates the NTPase of the nuclear periphery it is an essential component of the medium, but concentrations as high as 5 mM result in increased DNA leakage; there is also a slight increase in RNA release, which was overlooked in previous experiments, leading to an erroneous statement that increasing the Mg^{2+} concentration from 2.5 mM to 5 mM had no effect on RNA transport (Agutter 1980). It is possible that higher concentrations of this ion stimulate the endonucleases to which the DNA release has been provisionally attributed. In the presence of high concentrations of cytoplasmic protein, the Mg^{2+} activity might be lowered sufficiently to prevent such stimulation.

Clawson et al. (1980b) reported that more newly-synthesized DNA is released than bulk DNA. In view of the now considerable evidence that nascent DNA is associated with the matrix (Berezney, Coffey 1975; Smith, Berezney 1980), this suggests that the proposed endonucleases act on DNA close to the matrix fibrils. The different effects of spermidine, Ca^{2+}, Mn^{2+}, Mg^{2+} and phosphate on DNA leakage might therefore be interpretable in terms of the effects of these ions on matrix dynamics or on the matrix-DNA association.

The conclusion from these studies is that although, for the kinds of experiments on the mechanism of RNA transport that we have undertaken, it is appropriate to simplify medium C to medium C_2. Further simplification by omission of spermidine, Ca^{2+} or Mn^{2+}, with or even without an increase of the Mg^{2+} concentration, entails a loss of adequate stabilization and nuclear restriction.

Characterization of the Proteins Released

One prediction from the effect of spermidine on the release of DNA of "chromatosome" size from the nuclei is that less histone will be released into medium C_2 than into media A or B. The results summarized in Figure 2 confirm this prediction. No results for media C or C_1 are available. Because of the cytosol proteins in these media, studies of this kind would require the use of labeled nuclear proteins, and results from such experiments would be difficult to interpret because of uncertainties about the specific activities of individual polypeptides. In addition to the histones, proteins in the 34,000 and 36,000 molecular weight regions are also released in media A and B. These are likely to be the proteins associated with HnRNA rather than cytoplasmic messenger (Roy et al. 1979); so this observation is consistent with the hybridization studies (Table 4). On this argument, the persistence of these bands in medium C_2 is disturbing, and various modifications of this medium have been attempted to prevent the elution of these polypeptides from the nuclei. No modification has as yet been successful in this respect. Replacement of the Tris with N-2-hydroxyethylpiperazine-N'-2-ethanesulphonate (Kletzein 1980), 3-(N-morpholino)-sulphonate or triethanolamine makes no difference either to the proteins released or to the extent of DNA or RNA release. Omission of the sulphydryl reagent (2-mercaptoethanol or dithiothreitol) results in increased leakage of DNA, RNA and proteins. More encouraging is the observation that in medium C_2, a polypeptide of approximately 78,000 molecular weight, corresponding in size to a protein associated with cytoplasmic messenger (Barrieux et al. 1975), is released in the presence but not the absence of ATP.

Phase-Contrast Microscopic Studies

Comparison of the morphological changes of the nuclei during incubation with the amounts of various nuclear components released yielded little additional information. Increased DNA leakage (decreased stability) was associated with marked swelling and aggregation of the nuclei. Thus, massive lysis and aggregation occurred rapidly in medium B, in accordance with the observations of Goidl et al. (1975). In medium A, swelling and aggregation occurred more slowly, becoming marked only after 10 min incubation. In medium C and its variants no similar changes occurred, but after incubation for more than 30 min the nucleoplasm appeared more granular and there was significant shrinking of the nuclei. This alters the permeability properties of the nuclei, as the following experiment indicates. Nuclei were incubated in medium C_2 at 35°C for 30 min in the absence of

ATP. After this, 2.0 mM ATP was added. The ATP-stimulated release of RNA in the 8-18S molecular weight region after incubation for a further 20 min was not significant in This observation, that prolonged incubation in medium C and its variants irreversibly changes the properties of the nuclei, indicates that incubation times should, if possible, be limited to 20 min.

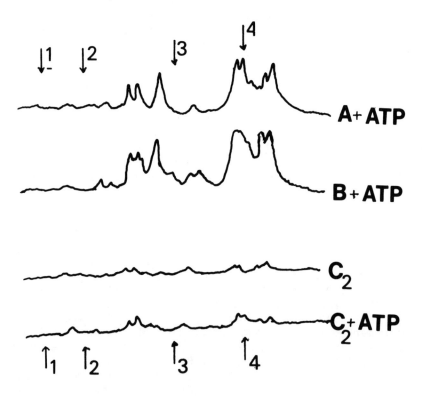

FIG 2

NUCLEAR PROTEINS RELEASED: 5 MIN INCUBATION

Fig. 2. Standards: 1: 168,000
2: 68,000
3: 23,000
4: 11,700

CONCLUDING REMARKS

So far as establishment of the mechanism of RNA transport is concerned, our studies overall have indicated that a simplified varient, called C_2 in this paper, of medium C (Yu et al. 1972) is the incubation system of choice for studies with isolated nuclei. That it is not ideal is indicated by the hybridization studies (Table 4) and protein release (Fig. 2), but these and other results show that it is superior to other alternatives in terms of the criteria adopted.

However, two caveats must be entered. First, results obtained by workers using other media can by no means be discounted. Thus, Clawson et al. (1980a, b) have obtained data that generally support our conclusions about the role of the NTPase of the nuclear periphery. Second, the experimental potential of medium C_2 is limited. Although its use has enabled us to clarify some important aspects of the mechanism, it has so far prevented any study of the complicated effects of the different cytosol components, and the importance of these is underlined by the observation of clear-cut differences between hepatoma and hepatocyte cytosols (Schumm et al. 1973a; Clawson et al. 1980a; Lemaire et al. 1981). Moreover, the possibility that certain cytosol components are essential for ribosome release (Racevskis, Webb 1974; Schumm et al. 1979) makes it unlikely that ribosome transport can usefully be studied with medium C_2. Nevertheless, the development and use of the simplest adequate system for a limited area of study goes some way towards ensuring that the results are not subject to interference by artefacts, and in addition reveals information about factors that affect the stability of isolated nuclei, which is likely to throw some incidental light on the mechanisms whereby normal nuclear structure is maintained.

ACKNOWLEDGEMENTS

I am indebted to the Cancer Research Campaign for supporting these investigations and for financing my attendance at this Symposium, and to Mr. C. D. Gleed for excellent technical assistance.

Agutter PS (1980). Influence of nucleotides, cations and nucleoside triphosphatase inhibitors on the release of ribonucleic acid from isolated rat liver nuclei. Biochem J 188:191.

Agutter PS (1979). Functional differences between mammalian nuclear matrix and pore-lamina preparations. Exp Cell Res 124:453.

Agutter PS, McArdle HJ, McCaldin B (1976). Evidence for involvement of nuclear envelope nucleoside triphosphatase in nucleo-cytoplasmic translocation of ribonucleoproteins. Nature 263:165.

Agutter PS, McCaldin B, McArdle HJ (1979). Importance of mammalian nuclear envelope nucleoside triphosphatase in nucleo-cytoplasmic transport of ribonucleoproteins. Biochem J 182:811.

Agutter PS, Ramsay I (1979). Further studies on the stimulation of the nuclear envelope nucleoside triphosphatase by polyguanylic acid. Biochem Soc Transac 7:720.

Agutter PS, Richardson JCW (1980). Nuclear nonchromatin proteinaceous structures: their role in the organization and function of the interphase nucleus. J Cell Sci 44:395.

Barrieux A, Ingraham HA, David DN, Rosenfeld MG (1975). Isolation of messenger-like ribonucleoproteins. Biochemistry 14:1815.

Berezney R (1979). Dynamic properties of the nuclear matrix. In Busch H (ed): "The Cell Nucleus", Vol VIII, New York: Academic Press, p 413.

Berezney R, Buchholtz LA (1978). Nuclear matrices isolated with protease inhibitors: evidence for an increased association with newly-replicated DNA. J Cell Biol 79:130a.

Berezney R, Coffey DS (1975). Nuclear protein matrix: association with newly synthesized DNA. Science 189:291.

Blobel G, Potter VR (1966). Nuclei from rat liver: isolation method that combined purity with high yield. Science 154:1662.

Burton K (1956). A study of the conditions and mechanism of the diphenylamine reaction for the colorimetric estimation of deoxyribonucleic acid. Biochem J 62:315.

Chatterjee NK, Weissbach H (1973). Release of RNA from HeLa cell nuclei. Arch Biochem Biophys 157:160.

Clawson GA, Koplitz M, Moody DE, Smuckler EA (1980a). Effects of thioacetamide treatment on nuclear envelope nucleoside triphosphatase activity and transport of RNA from rat liver nuclei. Cancer Res 40:75.

Clawson GA, James J, Woo C, Friend DJ, Moody DE, Smuckler EA (1980b). Pertinence of nuclear envelope nucleoside triphosphatase activity to ribonucleic acid transport. Biochemistry 19:2748.
Clawson GA, Smuckler EA (1980). Altered restriction of nuclear RNA during incubation in vitro. Biochem Biophys Res Commun 95:696.
Franke WW, Deumling B, Ermen B, Jarasch E-D, Kleinig H (1970). Nuclear membranes from mammalian liver. I: isolation procedure and general characterization. J Cell Biol 46:379.
Goidl JA, Canaani D, Boublik M, Weissbach H, Deckerman H (1975). Polyanion-induced release of polyribosomes from HeLa cell nuclei. J Biol Chem 250:9198.
Harris JR, Milne JF (1974). A rapid procedure for the isolation and purification of rat liver nuclear envelope. Biochem Soc Trans 2:1251.
Herlan G, Giese G, Wunderlich F (1979). Influence of nuclear membrane lipid fluidity on nuclear RNA release. Exp Cell Res 118:305.
Ishikawa K, Kuroda C, Ogata K (1969). Release of ribonucleoprotein particles containing rapidly labeled ribonucleic acid from rat liver nuclei. Effect of ATP on some properties of the particles. Biochim Biophys Acta 179:316.
Ishikawa K, Sato-Odani S, Ogata K (1978). The role of ATP in the transport of rapidly-labeled RNA from isolated nuclei of rat liver in vitro. Biochim Biophys Acta 521:650.
Kay RR, Fraser D, Johnston IR (1972). A method for the rapid isolation of nuclear membranes from rat liver. Eur J Biochem 30:145.
Kletzein RF (1980). Nucleo-cytoplasmic transport of RNA: effect of 3'-deoxy ATP on RNA release from isolated nuclei. Biochem J 192:753.
Lemaire M, Baeyens W, Baugnet-Mahieu L (1981). On the altered nucleocytoplasmic transport in vitro of rapidly labeled RNA, in the presence of cytosol or serum from tumor-bearing rats. Biomedicine 34:47.
Lichtenstein AV, Shapot VS (1976). A model for cytoplasm-governed gene regulation. Biochem J 159:783.
Long BH, Huang CY, Pogo AO (1979). Isolation and characterization of the nuclear matrix in Friend erythroleukaemia cells: chromatin and HnRNA interactions with the nuclear matrix. Cell 18:1079.
McDonald JR, Agutter PS (1980). The relationship between polyribonucleotide binding and the phosphorylation and

dephosphorylation of nuclear envelope protein. FEBS letters 116:145.

Miller TE, Huang CY, Pogo AO (1978). Rat liver nucler skeleton and small molecular weight RNA species. J Cell Biol 76:692.

Monneron A (1974). One-step isolation and characterization of nuclear membranes. Phil Trans Roy Soc London (B) 268:101.

Murty CN, Verney E, Sidransky H (1980). Effect of tryptophan on nuclear envelope nucleoside triphosphatase activity in rat liver. Proc Soc Exp Biol Med 163:155.

Nagel WC, Wunderlich F (1977). Effect of temperature on nuclear membranes and nucleo-cytoplasmic RNA transport in Tetrahymena grown at different temperatures. J Membr Biol 32:151.

Paine PE, Moore LC, Horowitz SB (1975). Nuclear envelope permeability. Nature, London 254:109.

Racevskis J, Webb TE (1974). Processing and release of ribosomal RNA from isolated nuclei: analysis of the ATP-dependence and cytosol-dependence. Eur J Biochem 49:93.

Roy RK, Lau AS, Munro HN, Baliga BS, Sarkar S (1979). Relationship of in vitro synthesized poly(A)-containing RNA from isolated rat liver nuclei: characterization of the ribonucleoprotein particles involved. Proc Natl Acad Sci USA 79:1751.

Sato T, Ishikawa K, Ogata K (1977). Factors causing release of ribosomal subunits from isolated nuclei of regenerating rat liver in vitro. Biochim Biophys Acta 474:549.

Schumm DE, Hanausek-Walaszek M, Yannarell A, Webb TE (1977). Changes in nucleo-cytoplasmic RNA transport incident to carcinogenesis. Eur J Cancer 13:139.

Schumm DE, Morris HP, Webb TE (1973a). Cytosol-modulated transport of messenger RNA from isolated nuclei. Cancer Res 33:1821.

Schumm DE, McNamara DJ, Webb TE (1973b). Cytoplasmic proteins regulating messenger RNA release from nuclei. Nature (New Biol) 245:201.

Schumm DE, Niemann MA, Palayoor T, Webb TE (1979). In vivo equivalence of a cell-free system from rat liver for ribosomal RNA processing and transport. J Biol Chem 254:12126.

Schumm DE, Webb TE (1974). The in vivo equivalence of a cell-free system for RNA processing and transport. Biochem Biophys Res Commun 58:354.

Schumm DE, Webb TE (1978). Effect of adenosine 3':5'-monophosphate and guanosine 3':5'-monophosphate on RNA release from isolated nuclei. J Biol Chem 253:8513.

Shearer RW, Smuckler EA (1972). Altered regulation of the transport of RNA from nucleus to cytoplasm in rat hepatoma cells. Cancer Res 32:339.

Smith HC, Berezney R (1980). DNA polymerase α is tightly bound to the nuclear matrix of actively replicating liver. Biochem Biophys Res Commun 97:1541.

Stuart SE, Clawson GA, Rottman FM, Patterson RJ (1977). RNA transport in isolated myeloma nuclei. Transport from membrane-denuded nuclei. J Cell Biol 72:57.

Yu L-C, Racevskis J, Webb TE (1972). Regulated transport of ribosomal subunits from regenerating rat liver nuclei in a cell-free system. Cancer Res 32:2314.

EFFECT OF TRYPTOPHAN ON NUCLEAR ENVELOPE IN RAT LIVER:
EVIDENCE FOR INCREASED NUCLEAR RNA RELEASE

Challakonda N. Murty, Ethel Verney and Herschel
Sidransky

Department of Pathology
The George Washington University
Medical Center, Washington, D.C. 20037

Among the essential amino acids, tryptophan has been demonstrated to play a special role in the regulation of hepatic protein metabolism. This special role of tryptophan was first revealed in studies conducted to examine the effect of amino acid supply on hepatic polyribosomes and protein synthesis in fasted animals. These studies reported that when a complete amino acid mixture devoid of one amino acid (arginine, histidine, isoleucine, leucine, lysine, methionine, phenylalanine, threonine, tryptophan, tyrosine, or valine) was tube-fed once to fasted animals, with a single exception, it initiated a stimulatory response in the liver (a shift toward heavier polyribosomes), similar to that obtained after a single feeding of the complete amino acid mixture (Sidransky et al. 1968; Pronezuk et al. 1968). The exception was the complete amino acid mixture with the deletion of tryptophan. In its absence the response by the liver was not a stimulatory one but remained similar to that found in the fasted control animals which contained relatively fewer heavy polyribosomes with a corresponding increase of monomers and dimers (Fleck et al. 1965; Wunner et al. 1966; Sidransky et al. 1968; Pronezuk et al. 1968). These findings suggested that tryptophan has a unique role in regulating hepatic protein metabolism. Further support came from our laboratory where we found that a single administration of tryptophan alone, but not of other amino acids, in concentrations used in the complete amino acid mixture, to fasted animals induced a rapid shift in hepatic polyribosomes towards heavier aggregates and enhanced in vitro and in vivo protein synthesis (Sidransky et al. 1968; Sidransky et al. 1971). Subsequent studies from other laboratories confirmed our findings that tryptophan when administered to fasted animals induced a stimulatory response involving the hepatic polyribosomes and protein synthesis (Cammarano et al. 1968; Rothschild et al. 1969; Oravec, Sourkes 1970).

Other studies from our laboratory have revealed another interesting effect of tryptophan on the liver during certain pathologic states whereby the hepatic protein synthesizing system became disturbed due to the administration of selected hepatotoxic agents. Thus, tryptophan has been found to have a preventive and/or curative effect upon the polyribosomes and protein synthesis in the livers of animals treated with hepatotoxic agents, such as ethionine (Sidransky et al. 1972), actinomycin D (Sidransky, Verney 1972), puromycin (Sarma et al. 1973), hypertonic NaCl (Sidransky et al. 1976) or CCl_4 (Sidransky et al. 1977). In addition, Rothschild et al. (1971), using the isolated perfused rabbit liver, demonstrated that the hepatic polyribosomal disaggregation induced with alcohol administration was corrected (reaggregated) when tryptophan was added. It was interesting to note that this improved (stimulatory) effect of tryptophan on hepatic protein synthesis in the experimental animal occurred even though the mechanisms by which the hepatotoxic agents acted were quite different.

Another interesting observation was made when tryptophan was administered to rats in which hepatic protein synthesis was already enhanced by other means, phenobarbital, cortisone acetate or force-feeding a threonine-devoid diet. In these instances the administration of tryptophan was found to enhance further the hepatic polyribosomal patterns (shift toward heavier aggregation) and protein synthesis in vitro (Sidransky, Verney 1977; Verney, Sidransky 1978). Thus, the different studies described above indicate a special role for tryptophan in producing a stimulatory effect on hepatic polyribosomes and protein synthesis in the livers of animals under a variety of conditions: animals that have been fasted; animals subjected to selected hepatotoxic agents, which caused diminished hepatic protein synthesis; and animals in which hepatic protein synthesis was already enhanced by selected agents or dietary manipulation.

MECHANISM OF ACTION

Although many studies have been concerned with elucidating the mechanism by which tryptophan stimulates hepatic protein synthesis, the complete answer is still not yet available. However, studies from other laboratories and from our laboratory have enabled us to develop a picture as to some of the mechanisms that are involved as a consequence of the administration of tryptophan.

Tryptophan as the Limiting Amino Acid

Initially, the special role of tryptophan in hepatic protein metabolism was attributed to the very low levels of free tryptophan in the liver and blood, especially after fasting (Munro 1968). Studies by Hori et al. (1967) and by Hunt et al. (1969), using reticulocytes, suggested that tryptophan tRNA may become limiting. Similarly, it was considered that tryptophan tRNA may become limiting in livers of fasted animals because of no dietary tryptophan intake and, therefore, would limit the rate of hepatic protein synthesis. However, data based on other studies (Sarma et al. 1971; Sidransky, Verney 1971; Sidransky et al. 1972) suggest that the action of tryptophan on hepatic polyribosomes and protein synthesis is more complex than merely raising the low levels of tryptophan in liver or blood. When tryptophan was administered to fed mice in which the hepatic and blood levels of tryptophan were much higher than in fasted mice, the hepatic polyribosomes shifted toward heavier aggregation similar to that which occurred when using fasted animals (Sarma et al. 1971). Also, Allen et al. (1969) reported no difference in the levels of tryptophanyl-tRNA in fasted and fed rats, suggesting that this was not rate-limiting in fasted animals. In addition, as described earlier, tryptophan has been effective in enhancing hepatic protein synthesis in animals treated with selected hepatotoxic agents where the tryptophan levels in blood and liver were not changed but were similar to that in control animals (unpublished observations).

Alteration in Translational Control

Early studies by Munro and co-workers (Fleck et al. 1965; Munro 1968) and by us (Sidransky et al. 1968;

Sidransky, Verney 1971), in which actinomycin D was administered prior to the administration of tryptophan, suggested that tryptophan could act at the translational level of control of hepatic protein synthesis in rats or mice. This was based upon data from experiments in which actinomycin D was administered first to inhibit RNA synthesis followed by the administration of tryptophan. Tryptophan was still found to have a stimulatory effect on hepatic polyribosomes and in vitro protein synthesis. More recent studies by Jorgensen and Majumdar (1976) revealed that when well-fed adrenalectomized rats were used, the administration of tryptophan after pretreatment with actinomycin D caused an increase in [^{3}H]leucine incorporation into ferritin, albumin, transferrin and fibrinogen, compared to that in water-fed controls. Without using the inhibitor, tryptophan caused a somewhat greater increase of incorporation into the same components.

Alteration in Transcriptional Control

Other findings suggested that tryptophan administration may act at the transcriptional level of control of hepatic protein synthesis. Tryptophan administration has been reported to cause an increase in the liver of DNA-dependent RNA polymerase activity (Henderson 1970; Vesley, Cihak 1970), nuclear RNA synthesis (Henderson 1970; Vesley, Cihak 1970; Oravec, Korner 1971), polyribosomal RNA synthesis (Wunner 1967; Henderson 1970) and of cytoplasmic mRNA (Murty, Sidransky 1972; Murty et al. 1976).

Alteration in Nucleocytoplasmic Translocation of mRNA

Recent reports have demonstrated that tryptophan has an effect at the post-transcriptional level of control of hepatic protein synthesis. This conclusion was based upon several studies from our laboratory. After first observing that there was an increase in hepatic cytoplasmic mRNA following tryptophan administration (Murty, Sidransky 1972; Murty et al. 1976), we next explored whether this increase was due to increased synthesis or to increased nucleocytoplasmic transport of mRNA. In experiments where hepatic RNA of fasted mice was prelabeled with [^{14}C-6]orotic acid and then the animals were treated with actinomycin D to inhibit further RNA synthesis before being tube-fed tryptophan, we found elevated levels of cytoplasmic mRNA and a shift in polyribosomes toward heavier aggregates in the livers due to tryptophan (Murty, Sidransky 1972). This suggested that tryptophan may act to stimulate the transfer of mRNA into the cytoplasm of hepatic cells. In further studies we observed that the administration of tryptophan to fasted animals resulted in significant increases in the amounts of hepatic poly(adenylic acid) [poly(A)] and poly(A)-mRNA in the cytoplasm (Murty et al. 1976). Next, we explored whether this stimulation would occur following the administration of cordycepin [an inhibitor of poly(A) synthesis] and/or actinomycin D (an inhibitor of RNA synthesis). Administration of tryptophan to fasted rats pretreated with cordycepin or actinomycin D, or both, induced a shift in hepatic polyribosomes toward heavier aggregates and an increase in in vitro protein synthesis. Also, fasted rats that received [U-^{14}C]adenosine to prelabel hepatic poly(A) and then were treated with cordycepin, or actinomycin D, or both, before tube-feeding tryptophan revealed increased hepatic levels of labeled polyribosomal poly(A) in comparison with controls (Murty et al. 1976). In further studies we reported that the administration of tryptophan to fasted rats pretreated with cordycepin and actinomycin D led to decreased levels of nuclear poly(A)-mRNA and a concomitant increase in the levels of polyribosomal poly(A)-mRNA in the cytoplasm as determined by measuring in vivo incorporation of labeled precursors into hepatic RNA (Murty et al. 1977). These findings suggested that tryptophan plays a role in stimulating the rate of translocation of poly(A)-mRNA into the cytoplasm of the livers of normal animals (Murty, Sidransky 1972; Murty et al. 1976) and in those treated with selected inhibitors (Sidransky et al. 1976; Sidransky et al. 1977).

More recent studies have dealt with further aspects relating to the enhanced hepatic nucleocytoplasmic translocation of mRNA occurring following tryptophan administration. Evidence is accumulating that post-transcriptional controls are operating within the nucleus, at the nuclear membrane, or in the cytoplasm to regulate the rate of nucleocytoplasmic translocation of ribonucleoprotein particles both qualitatively and quantitatively (Brawerman 1974; Perry 1976). A number of investigators have reported evidence for modification of nucleocytoplasmic transport of RNA by chemical carcinogens (Smuckler, Koplitz 1976; Clawson et al. 1980), in transformed cells (Shearer 1979; Patel et al. 1979), and during aging (Yannarell et al. 1977). In view of the various post-transcriptional controls present in the nucleus as well as in the cytoplasm, we have undertaken studies in order to investigate the role of the nucleus, the cytoplasm, or both in the enhanced hepatic nucleocytoplasmic translocation of mRNA following tryptophan administration. Experiments were conducted using a cell-free system in which the release of RNA from isolated nuclei into a defined medium could be studied. This experimental approach was modeled after that used by Schumm and Webb (1974), who observed that the release of RNA from isolated hepatic nuclei was dependent upon dialyzed cell sap and suggested that the cell sap contained nondialyzable components that act post-transcriptionally to regulate the nuclear processing and/or transport of mRNA to the cytoplasm. Using this assay system, we obtained evidence indicating that tryptophan influenced both the nucleus and the cytoplasm in their regulation of intracellular transport of RNA. First, using control liver cell sap, we found that there was greater release of labeled poly(A)-mRNA (prelabeled in vivo with [^{14}C]orotic acid) into the medium from isolated hepatic nuclei of tryptophan-treated rats than from nuclei of control animals (Murty et al. 1977). This effect was detected within 10 min after the administration of tryptophan. Therefore, we investigated how early certain changes occurred after tryptophan administration in order to determine whether they may be involved or related to the overall effect. We observed that tryptophan concentrations became elevated in the plasma, in the liver and in hepatic nuclei within 5 min (Murty et al. 1979) and that increased hepatic protein synthesis was evident within 10 min (Sidransky et al. 1968). We also found that the hepatic nuclei of rats receiving tryptophan for as short an interval as 3 or 6 min demonstrated greater release of labeled RNA in vitro than did those from hepatic nuclei of control animals (Murty et al. 1979).

Since liver cell sap was found to be essential for the controlled release of mRNA from isolated hepatic nuclei, we investigated whether components in the liver sap, particularly proteins, of the in vivo tryptophan-treated animals were influential. This was of special interest since we had observed that in vivo hepatic protein synthesis was already stimulated within 10 min after tryptophan administration (Sidransky et al. 1968). Thus, conceivably some regulatory protein might play a role in the transport activity. To test for this possibility, cell saps were prepared from livers of rats tube-fed water or tryptophan 10 min before killing. Their effects on in vitro release of labeled poly(A)-mRNA from control hepatic nuclei in the presence of added tryptophan in the medium were investigated. The results revealed that there was more release of total RNA and poly(A)-mRNA from control hepatic nuclei incubated with liver sap of tryptophan-treated rats compared to release using liver sap of control rats (Murty et al. 1979). The results also revealed that the addition of tryptophan in the incubation medium was essential for the increased release of labeled RNA by the experimental liver sap. This addition was probably necessary due to the lack of tryptophan in the cell sap, since dialyzed cell saps were used in the incubation medium.

Since hepatic protein synthesis was rapidly stimulated following tryptophan administration, we next examined whether pretreatment of animals with inhibitors of protein synthesis would alter the increased transport activity of liver sap of tryptophan-treated rats. Animals were pretreated with cycloheximide for 2 1/2 hr or with puromycin for 20 min prior to tryptophan or water administration given 10 min before killing. Liver saps were prepared from control and experimental groups, and their effects on in vitro release of labeled RNA from hepatic nuclei were investigated. Liver saps from rats treated with cycloheximide or puromycin prior to tryptophan administration were not able to stimulate the release of labeled RNA as was the case with liver saps of tryptophan-treated rats (Murty et al. 1979).

In earlier studies of animals treated with inhibitors (those primarily inhibiting protein synthesis, such as ethionine, puromycin, hypertonic NaCl or CCl_4) and which received tryptophan, revealed that tryptophan was still able to induce a stimulatory effect on hepatic protein synthesis and polyribosomal aggregation (Sarma et al. 1973; Sidransky et al. 1972; Sidransky et al. 1976; Sidransky et al. 1977). However, the results described above indicated that the liver cell sap factor may not be involved in this stimulatory effect by tryptophan, since the enhancing effect of this factor on the flow of RNA from nuclei into the medium was lost by pretreatment with puromycin or cycloheximide. Therefore, we considered that the ability of tryptophan to produce the stimulatory effect in animals treated with inhibitors of protein synthesis may be related to enhanced nucleocytoplasmic translocation of mRNA, probably through altered transport activity of hepatic nuclei itself without the cell sap effect. To test for this possibility, nuclei were isolated from livers of rats that received puromycin followed by tryptophan administration and the isolated nuclei were examined for in vitro release of RNA into the medium. Hepatic nuclei of rats that received tryptophan following puromycin exhibited greater release of labeled RNA [poly(A)-mRNA] into the medium in comparison to control of puromycin-treated liver nuclei (Murty et al. 1979).

Involvement of the Hepatic Nuclear Envelope

A more recent finding has cast new light in regard to the enhanced hepatic nucleocytoplasmic translocation of RNA in tryptophan-treated rats. A number of studies suggest that the nuclear envelope and the nuclear pore complex, in particular, play a key role in the regulation of nucleocytoplasmic RNA translocation (Harris 1978). A nucleoside triphosphatase (NTPase) has been identified in the mammalian liver nuclear envelope and this enzyme appears to be involved in the nucleocytoplasmic translocation of RNA (Clawson et al. 1980; Agutter et al. 1979; Vorbrodt, Maul 1980). Furthermore, following treatment of rats with thioacetamide or CCl_4, a parallelism between alterations in nuclear RNA transport and nuclear envelope NTPase activity in the liver was demonstrated (Clawson et al. 1980). Therefore, we investigated the possibility that the enhanced nucleocytoplasmic translocation of mRNA due to tryptophan was related to an alteration in the activity of nuclear envelope NTPase. The results presented in Figure 1A demonstrate that the levels of nuclear envelope NTPase activity were significantly elevated in the livers of rats tube-fed tryptophan 10, 30 or 60 min before killing (Murty et al. 1980). As described earlier and shown in Figure 1B, concomitant with this rapid (10 min) increase in the NTPase activity, there was more release of labeled RNA from isolated hepatic nuclei of livers of tryptophan-treated rats than from those of control rats. The parallel increases in both NTPase activity of nuclear envelopes and the translocation of RNA suggest that these two processes may be associated with or related to one another.

In view of the probable importance of the nuclear envelope in controlling active nucleocytoplasmic transfer of RNP particles, several workers have examined the ultrastructure

EFFECT OF TRYPTOPHAN ON NUCLEAR−ENVELOPE NTPASE ACTIVITY AND ON *IN VITRO* RELEASE OF LABELED RNA FROM ISOLATED LIVER NUCLEI

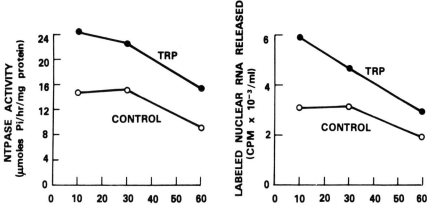

TIME OF ADMINISTRATION (MIN.)

of the nuclear envelope. The intact nuclear envelope is composed of inner and outer nuclear membranes and nuclear pore complexes (Aaronson, Blobel 1974). Treatment of nuclei with non-ionic detergents, such as Triton X-100, completely removes the outer nuclear membrane, leaving intact nuclei with preservation of nuclear pore complexes. Evidence has been presented suggesting that the nuclear pores are the major sites of nucleocytoplasmic transfer of macromolecules (Franke 1974; Stuart et al. 1977). Therefore, we examined the effect of removal of outer nuclear membrane by Triton X-100 treatment of isolated hepatic nuclei of control and tryptophan-treated rats on their capacity to transport RNA in vitro and on nuclear envelope NTPase activity. Following treatment with Triton X-100, an increased release of labeled RNA persisted from the hepatic nuclei of tryptophan-treated rats compared with that from control hepatic nuclei (Murty et al. 1980). These findings were similar to those observed with nuclei not subjected to Triton X-100 treatment. Similar increases in the activity of nuclear envelope NTPase were found with the experimental compared with the control samples (Murty et al. 1980). In addition to our finding that increased translocation of mRNA was probably related to increased levels of nuclear envelope NTPase activity in the livers of the experimental rats, these data are in general agreement with those of others in that: (a) detergent treatment was not deleterious on the ability to transport RNA, (b) the transported RNA was of the intranuclear origin, and (c) nuclear pore complexes and NTPase activity were probably responsible for the nucleocytoplasmic translocation of mRNA (Agutter et al. 1979; Stuart et al. 1977; Palayoor et al. 1981).

In further studies we also found that the administration of tryptophan was able to stimulate the levels of hepatic nuclear envelope NTPase activity in rats pretreated with puromycin, similar to the increases in the control rats that received tryptophan alone (Murty et al. 1980). This again suggested that increased activity of nuclear envelope NTPase was probably responsible for the enhanced RNA transport activity of hepatic nuclei of puromycin plus tryptophan-treated rats compared to that of the hepatic nuclei of control rats.

Table 1 summarizes our current views in regard to explaining the increased cytoplasmic mRNA in liver after tryptophan administration. Figure 2 presents a schematic diagram as to how we currently envision the effects of tryptophan on hepatic protein synthesis. Our current thoughts are that tryptophan can act to stimulate liver protein synthesis in one or both of two ways: (1) enhancement of mRNA synthesis and (2) increased nucleocytoplasmic translocation of mRNA. In normal animals, tryptophan probably acts by both mechanisms. In animals treated with inhibitors of RNA metabolism, such as cordycepin and actinomycin D, it acts only by the second mechanism. In animals treated with other inhibitors, those which primarily inhibit protein synthesis such as ethionine, puromycin or CCl_4, the second mechanism may also be mainly involved, but a full explanation is not yet available. Concerning the phenomenon of nucleocytoplasmic translocation of mRNA, it is not yet understood how the post-transcriptional controls located within the nucleus and cytoplasm regulate the flow of RNA from nucleus to cytoplasm. Our studies suggest that in normal animals, tryptophan affects both the nucleus as well as the cytosol in leading to an increase in the levels of cytoplasmic mRNA. The mechanism by which this occurs may involve an actively synthesized or turning over protein which regulates or controls nucleocytoplasmic flow of mRNA. Our findings suggest that this protein may be at least in part an enzyme, NTPase, which may be related to or responsible for the transportation of mRNA across the nuclear envelope. Although this

TABLE I

Explanation for Increased Cytoplasmic mRNA in Liver After Tryptophan Administration

---

I. Increased Synthesis of mRNA

    1. Increased DNA-dependent RNA polymerase activity
    2. Increased DNA-like RNA synthesis

II. Increased Translocation of mRNA

    1. Increased cytoplasmic mRNA (after actinomycin D)
    2. Increased cytoplasmic poly(A) and poly(A)-mRNA (after cordycepin and actinomycin D)
    3. Increased levels of NTPase activity in nuclear envelopes (after actinomycin D and puromycin).
    4. Increased release of mRNA from nucleus to cytoplasm (in vivo and in vitro) (nuclear and cytosol effects)
    5. Increased release of mRNA from nucleus to cytoplasm after puromycin (nuclear effect)

III. Both I and II

    1. In normal animals, I and II may act
    2. In animals treated with inhibitors, only II may act

---

enzyme has been characterized with respect to its substrate specificity and kinetic behavior, little is known concerning its synthesis and turnover. Since there is already marked enhancement of in vivo incorporation of labeled leucine into proteins of nucleus, nuclear envelope and cytoplasm within 10 min following the administration of tryptophan (unreported data), it is possible that this change may be related to the rapid synthesis of one or several specific regulatory proteins, possibly such as NTPase of nuclear envelope, resulting in greater outpouring of nuclear RNA into cytoplasm. Protein factors in the nucleus and cytosol of rat liver may act as important regulators of nuclear transport activity and may be altered by a variety of factors or means, including nutritional components and drugs.

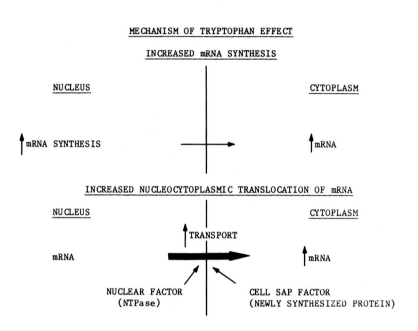

FIGURE 2

Recently another enzyme, protein phosphokinase, has been identified on the rat liver nuclear envelope and it has been suggested that this enzyme may also regulate nuclear transport activity via the phosphorylation and dephosphorylation of specific regulatory proteins (Steer et al. 1979). Results of our preliminary experiments indicate that the protein phosphokinase activity of rat liver nuclear envelope was enhanced following the administration of tryptophan.

Other mechanisms may also be involved in the enhanced nuclear RNA transport activity due to tryptophan. Schumm and Webb (1978) have reported that the addition to cyclic AMP or GMP stimulated the release of RNA from isolated hepatic nuclei, indicating that cyclic nucleotides can exert an influence on the post-transcriptional events of RNA processing and transport. Using this approach in preliminary experiments, we have found that the addition of cAMP to the cell-free system with cells saps of livers of control rats but not of livers of tryptophan-treated rats caused increases in the release of labeled RNA from control rats liver nuclei. This response in the cell-free system to added cAMP reflects the in vivo concentrations in the tissues from which the system was prepared. Thus, our findings suggest that tryptophan may elevate in vivo the cAMP levels in liver cytosol and this may be of importance in the enhanced nucleocytoplasmic translocation of mRNA in liver due to tryptophan. Further studies are being conducted to determine whether this is indeed the case.

Another interesting aspect of our present study is that tryptophan could affect the hepatic nuclei independent of new protein synthesis. Although the activity of the cytosol factor was lost by pretreatment with puromycin, the hepatic nuclei of puromycin and tryptophan-treated rats exhibited enhanced RNA transport activity. At present is is not known as to how under these conditions tryptophan influences the nuclei alone to enhance nucleocytoplasmic translocation of mRNA. It is possible that in animals treated with inhibitors of RNA or protein synthesis, tryptophan may either stimulate the synthesis of regulatory proteins in the nucleus or may stimulate the transport of regulatory factors from the cytosol to the nuclear membrane. Although the question of whether nuclear protein synthesis occurs is as yet not answered, some investigators have implicated the nucleus and nuclear envelope as sites of protein synthesis and indicated that nuclear protein synthesis is inhibited to a lesser extent by puromycin and cycloheximide than is cytoplasmic protein synthesis (Ono et al. 1976; Ono, Terayama 1968). Thus, it becomes quite difficult to unravel the complex mechanisms which come into play following the use of tryptophan singly or in combination of a variety of inhibitors. Indeed, inhibitors which are considered to have one primary action often cause a variety of secondary effects in vivo in animals. Although the use of inhibitors of RNA and protein synthesis are of great value in unraveling mechanisms, caution must be exercised in interpreting their overall effects when relating to mechanisms in normal animals.

Tryptophan appears to be a unique amino acid which rapidly stimulates hepatic protein synthesis in normal animals and also in animals undergoing certain pathologic changes. This paper has reviewed some of the current experimental data and has attempted to elucidate how tryptophan is currently thought to act. An interesting picture is developing as to the mechanisms that may be involved. Further studies are needed to complete the picture.

ACKNOWLEDGMENT

This investigation was supported by U.S. Public Health Service Research Grants AM-27339 from the National Institute of Arthritis, Metabolism and Digestive Diseases and CA-26557 from the National Cancer Institute.

Aaronson RP, Blobel G (1974). On the attachment of the nuclear pore complex. J Cell Biol 62:746.
Agutter P, McCaldin B, McArdle HJ (1979). Importance of mammalian nuclear-envelope nucleoside triphosphatase in nucleocytoplasmic transport of ribonucleoproteins. Biochem J 182:811.
Allen RE, Raines PL, Regen DM (1969). Regulatory significance of transfer RNA changing levels. I. Measurements of changing levels in livers of chow-fed rats, fasting rats, and rats fed balanced or imbalanced mixtures of amino acids. Biochim Biophys Acta 190:323.
Brawerman G (1974). Eukaryotic messenger RNA. Ann Rev Biochem 43:621
Cammarano P, Chinali G, Gaetami S, Spadoni MA (1968). Involvement of adrenal steroids in the changes of polysome organization during feeding of imbalanced amino acid diets. Biochim Biophys Acta 155:302.
Clawson GA, James J, Woo CH, Friend DS, Moody D, Smuckler EA (1980). Pertinence of nuclear envelope nucleoside triphosphatase activity to ribonucleic acid transport. Biochemistry 19:2748.
Clawson GA, Woo CH, Smuckler EA (1980). Independent responses of nucleoside triphosphatase and protein kinase activities in nuclear envelope following thioacetamide treatment. Biochem Biophys Res Commun 95:1200.
Fleck A, Shepherd J, Munro HN (1965). Protein synthesis in rat liver. Influence of amino acids in diet on microsomes and polysomes. Science 150:628.
Franke WW (1974). Structure, biochemistry and functions of the nuclear envelope. Int Rev Cytol 4:71.

Harris JR (1978). The biochemistry and ultrastructure of the nuclear envelope. Biochim Biophys Acta 515:55.

Henderson AR (1970). The effect of feeding with a tryptophan-free amino acid mixture on rat liver magnesium ion-activated deoxyribonucleic acid-dependent ribonucleic acid-dependent polymerase. Biochem J 120:205.

Hori M, Fisher JM, Robinovitz M (1967). Tryptophan deficiency in rabbit reticulocytes: polyribosomes during interrupted growth of hemoglobin chains. Science 155:83.

Hunt RT, Hunter AR, Munro AJ (1969). The control of hemoglobin synthesis: factors controlling the output of alpha and beta chains. Proc Nutr Soc Engl Scot 28:248.

Jorgensen AJF, Majumdar APN (1976). Bilateral adrenalectomy: effect of tryptophan force-feeding on amino acid incorporation into ferritin, transferrin, and mixed proteins of liver, brain and kidneys in vivo. Biochem Med 16:37.

Munro HN (1968). Role of amino acid supply in regulating ribosome function. Fed Proc 27:1231.

Murty, CN, Sidransky H (1972). The effect of tryptophan on messenger RNA of the livers of fasted mice. Biochim Biophys Acta 262:328.

Murty, CN, Verney E, Sidransky H (1976). Effect of tryptophan on polyriboadenylic acid and polyadenylic acid-messenger ribonucleic acid in rat liver. Lab Invest 34:77.

Murty, CN, Verney E, Sidransky H (1977). The effect of tryptophan on nucleocytoplasmic translocation of RNA in rat liver. Biochim Biophys Acta 474:117.

Murty, CN, Verney E, Sidransky H (1979). In vivo and in vitro studies on the effect of tryptophan on translocation of RNA from nuclei of rat liver. Biochem Med 22:98.

Murty, CN, Verney E, Sidransky H (1980). Effect of tryptophan on nuclear envelope nucleoside triphosphatase activity in rat liver. Proc Soc Exp Biol Med 163:155.

Ono H, Ono T, Wada O (1976). Amino acid incorporation by nuclear membrane fraction of rat liver. Life Sci 18:215.

Ono H, Terayama H (1968). Amino acid incorporation into proteins in isolated rat liver nuclei. Biochim Biophys Acta 166:175.

Oravec M, Korner A (1971). Stimulation of ribosomal and DNA-like RNA synthesis by tryptophan. Biochim Biophys Acta 247:404.

Oravec M, Sourkes TL (1970). Inhibition of hepatic protein synthesis by α-methyl-DL-tryptophan in vivo. Further studies on the glyconeogenic action of α-methyl-tryptophan.

Palayoor T, Schumm DE, Webb TE (1981). Transport of functional messenger RNA from liver nuclei in a reconstituted cell-free system. Biochim Biophys Acta 654:201.

Patel NT, Folse DS, Holoubek V (1979). Release of repetitive nuclear RNA into the cytoplasm in liver of rats fed 3'-methyl-4-dimethylaminoazobenzene. Cancer Res 39:4460.

Perry RP (1976). Processing of RNA. Ann Rev Biochem 45:605.

Pronezuk AW, Baliga BS, Triant JW, Munro HN (1968). Comparison of the effect of amino acid supply on hepatic polysome profiles in vivo and in vitro. Biochim Biophys Acta 157:204.

Rothschild MA, Oratz M, Mongelli J, Fishman L, Schreiber SS (1969). Amino acid regulation of albumin synthesis. J Nutr 98:395.

Rothschild MA, Ortaz M, Mongelli J, Schreiber SS (1971). Alcohol-induced depression of albumin synthesis: reversal by tryptophan. J Clin Invest 50:1812.

Sarma DSR, Bongiorno M, Verney E, Sidransky H (1973). Effect of oral administration of tryptophan or water on hepatic polyribosomal disaggregation due to puromycin. Exp Mol Path 19:23.

Sarma DSR, Verney E, Bongiorno M, Sidransky H (1971). Influence of tryptophan on hepatic polyribosomes and protein synthesis in non-fasted and fasted mice. Nutr Rep Inter 4:1.

Schumm DE, Webb TE (1974). Modified messenger ribonucleic acid release from isolated hepatic nuclei after inhibition of polyadenylate formation. Biochem J 139:191.

Schumm DE, Webb TE (1978). Effect of adenosine 3':5' monophosphate and guanosine 3':5'-monophosphate on RNA release from isolated nuclei. J Biol Chem 253:8513.

Shearer RW (1979). Altered RNA transport without derepression in rat kidney tumor induced by dimethylnitrosamine. Chem Biol Inter 27:91.

Sidransky H, Sarma DSR, Bongiorno M, Verney E (1968). Effect of dietary tryptophan on hepatic polyribosomes and protein synthesis in fasted mice. J Biol Chem 243:1123.

Sidransky H, Verney E (1972). Effect of diet and tryptophan on hepatic polyribosomal disaggregation due to actinomycin D. Exp Mol Path 17:233.

Sidransky H, Verney E (1977). Effect of tryptophan or phenobarbital administration on hepatic polyribosomes and protein synthesis in rats force-fed a complete or threonine-devoid diet. J Nutr 107:730.

Sidransky H, Verney E, Murty CN (1976). Effect of tryptophan on hepatic polyribosomal disaggregation due to hypertonic sodium chloride. Lab Invest 34:291.

Sidransky H, Verney E, Murty CN (1977). Effect of tryptophan on hepatic polyribosomes and protein synthesis in rats treated with carbon tetrachloride. Tox Applied Pharm 39:295.

Sidransky H, Verney E, Sarma DSR (1971). Effect of tryptophan on polyribosomes and protein synthesis in liver. Am J Clin Nutr 24:779.

Sidransky H, Verney E, Sarma DSR (1972). Effect of tryptophan on hepatic polyribosomal disaggregation due to ethionine. Proc Soc Exp Biol Med 149:633.

Smuckler EA, Koplitz RM (1976). Polyadenylic acid content and electrophoretic behavior of in vitro released RNA's in chemical carcinogenesis. Cancer Res 36:881.

Steer RC, Wilson MJ, Ahmed K (1979). Protein phosphokinase activity of rat liver nuclear membranes. Exp Cell Res 119:403.

Stuart SE, Clawson GA, Rottman FM, Patterson RJ (1977). RNA transport in isolated myeloma nuclei. Transport from membrane-denuded nuclei. J Cell Biol 72:57.

Verney E, Sidransky H (1978). Further enhancement by tryptophan of hepatic protein synthesis stimulated by phenobarbital or cortisone acetate. Proc Soc Exp Biol Med 158:245.

Vesley J, Cihak A (1970). Enhanced DNA-dependent RNA polymerase and RNA synthesis in rat liver nuclei after administration of L-tryptophan. Biochim Biophys Acta 204:614.

Vorbrodt A, Maul GG (1980). Cytochemical studies on the relation of nucleoside triphosphatase activity to ribonucleoproteins in isolated rat liver nuclei. J Histochem Cytochem 28:28.

Wunner WH (1967). The time sequence of RNA and protein synthesis in cellular compartments following acute dietary challenge with amino acid mixtures. Proc Nutr Soc 27:153.

Wunner WH, Bell J, Munro HN (1966). The effect of feeding with a tryptophan-free amino acid mixture on rat liver polysomes and ribosomal ribonucleic acid. Biochem J 101:417.

Yannarell A, Schumm DE, Webb TE (1977). Age-dependence of nuclear RNA processing. Mech Aging Dev 6:259.

IMMUNOLOGICAL PROBES TO INVESTIGATE THE ROLE OF THE NUCLEAR ENVELOPE IN RNA TRANSPORT

F. A. Baglia and G. G. Maul

The Wistar Institute of Anatomy and Biology
36th Street at Spruce
Philadelphia, PA 19104

The nuclear envelope is a partition that separates the major compartments of the cell. The significance of this membranous partition in the exchange of macromolecules between the transcriptional and translational compartments remains unclear. Undoubtedly, the nuclear envelope could regulate such cellular processes as protein synthesis and gene expression (by controlling RNP transport). An in-depth investigation of the nuclear envelope at the molecular level would clarify the importance of nucleocytoplasmic exchange.

Several studies suggest that exchange between the two major compartments is facilitated by nuclear pore complexes (Maul 1977; Feldherr 1972; Franke, Scheer 1974) and those processes involving active transport are involved in the movement of ribonucleoproteins at the pore complex site (Feldherr 1972; Agutter 1972). These complexes have been well described ultrastructurally (Maul 1977), though the chemical features of the "pores" remain uncharacterized. Evidence for the movement of macromolecules through the pore complex has also come from ultrastructural studies in which large RNP aggregates were visualized at the site of the pore complex (Franke, Scheer 1974; Maul 1977).

A cell-free system has been utilized to investigate the transport of RNA (Schumm, Webb 1974), but the components of the nuclear envelope that are involved in transport and/or processing of RNA have not been investigated directly. Experiments in which RNA transport from rat liver nuclei and from detergent-treated nuclei was compared suggest that removal of nuclear membranes does not affect RNA transport

(Steward et al. 1977), thus implicating the pore complex as the site of RNA transport and/or processing. The proteins that comprise the nuclear pore complex as well as the polypeptides involved in the transport of RNP have not been identified.

We have prepared immunological reagents directed against clam nuclear envelopes and against individual envelope proteins and groups of proteins. Germinal vesicles (GV) from the surf clam Spisula solidissima were used as a source of nuclear envelope in our study. The GV envelope contains the pore complex in the highest possible packing ratio (Maul 1980). Furthermore, enzyme or high-salt extraction methods are not necessary for their isolation, eliminating the possible removal of weakly attached, essential components.

We utilized a cell-free RNA transport system to screen for a population of antibodies that blocks the efflux of RNP, in an attempt to determine the antigen(s) that comprise the RNP transport system. In our initial experiments, we raised antibodies in rabbits to clam nuclear envelopes to establish whether the binding of immunoglobulin to nuclear envelope antigens inhibits ATP-dependent RNP release. Isolated rat liver nuclei that had been radiolabeled with [^{14}C]orotic acid were used to measure ATP-dependent RNP release. In our hands, ATP stimulated the release of RNP to a level at least twice that without ATP. Incubation of nuclei with preimmune IgG had no effect on the amount of RNP released. However, antibodies directed against clam nuclear envelopes substantially reduced ATP-dependent RNP release (Table 1). These findings allowed us to identify the antigen(s) that regulates the release of RNP. Nuclear envelopes extracted with 1 M NaCl and 1% Triton X-100 were separated by sodium dodecyl sulfate-polyacrylamide gel electrophoresis. Different protein fractions were eluted from acrylamide gels and crosslinked with glutaraldehyde (Fuller et al. 1975). The emulsified antigens (Freund's adjuvant incomplete:complete, 8:1) were injected into white Leghorn chickens. The antibody response was detected by indirect immunofluorescence. Antibodies directed against protein(s) 1 and 67K gave an intense nuclear fluorescent staining in an endothelial cell line (bovine fetal aortic cells). Antibodies directed against the other antigens also showed nuclear staining, though only antibodies against the 190K antigen demonstrated weak cytoplasmic staining. Only antibodies directed against protein(s) 1 reduced RNP release (Table 1). Thus, a population of antibodies directed

Table 1
Reactivity of Antibodies Against Clam Nuclear Envelopes or Envelope Proteins on ATP-dependent Release of RNA from Isolated Rat Liver

	Cpm/1.5×10^6 nuclei
Control (no IgG) + ATP	4479
Control (no IgG) − ATP	2196
Preimmune IgG + ATP	4759
Preimmune IgG − ATP	2382
Anti-nuclear envelope + ATP	3526
(anti-1) + ATP	2021
(anti-2) + ATP	4550
(anti-3) + ATP	4435
(anti-4) + ATP	4746
(anti-190K) + ATP	4440
(anti-6) + ATP	4685
(anti-67K) + ATP	4518
Cationized ferritin 1 mg/ml	2487

against proteins in a high-molecular-weight class effectively reduced ATP-dependent RNP release. Apparently, this class of proteins is involved in regulating the translocation of RNP in intact nuclei.

To determine whether anti-1 blocked ATP-dependent RNA transport by reacting with membranes (e.g. lipoproteins) or by simply precipitating RNP in the nuclei, we tested the specificity of the antibodies. Even for incubation periods longer than those for intact nuclei, anti-1 did not precipitate RNP. Thus, anti-1 does not crossreact immunologically with RNP, nor does it inhibit ATP-dependent RNP release by precipitating nuclear RNP. These results were confirmed in radioimmunoassay. Additional experiments determined that anti-1 does not crossreact with rRNA, calf-thymus DNA or rat liver microsomes, suggesting that anti-1 does not contain an antibody population directed against rRNA, DNA or membrane components that may be involved in transport at the pore complex site.

To examine the possibility that anti-1 could be crosslinking proteins at or near the pore complex and thus sterically block RNP release, we used cationized ferritin, a highly charged molecule which attaches to all surfaces, including the pore complex. This molecule was found to inhibit ATP-dependent RNP release (Table 1).

Radioimmunoassay demonstrated that clam anti-1 is crossreactive with rat fibrous lamina pore complex (FLPC) fractions (Table 2). After dissolving rat FLPC complex in 5 M urea and 10 mM DTT and placing the mixture on microtiter plates, crossreactivity with low dilutions of antigen could be detected using radiolabeled rabbit anti-chicken IgG. These results, together with those of the indirect immuno-

Table 2

Radioimmunoassay of a Rat Liver Nuclear Envelope Fraction with Clam Anti-1

	Clam FLCP	Rat FLPC	No antigen
Anti-1 control, IgG	207	198	
Anti-1 (antigen)	2463	855	218

fluorescence assay, indicate that protein(s) 1 must be highly conserved, as anti-1 is crossreactive with two phylogenetically distant species (clam and rat).

We attempted to localize the structural components to which anti-1 is directed by indirect immunoperoxidase staining of clam nuclei. Reaction products appeared throughout the nucleus, chromosomes and particularly on the outer nuclear envelope of clam nuclei, obscuring the pore complex.

We are currently attempting to characterize our transport blocking antibody. Because NTPase activity is correlated with RNP translocation in this cell-free system (Agutter 1976), it is possible that anti-1 is directed against an active component of this enzyme. On the other hand, anti-1 might crosslink the structural proteins that comprise the pore complex and in this way block the efflux of RNP. Unravelling the molecular structure of the pore complex should be a challenging task and awaits the isolation and characterization of these high-molecular-weight proteins.

Agutter PS, McArdle HY, McCaldin B (1976). Nature (London) 263:165.
Feldherr CM (1972). Adv Cell Mol Biol 2:273.
Franke WW, Scheer U (1974). In Bush H (ed): "The Cell Nucleus", Vol 1, New York: Academic Press, p 273.
Fuller GM, Brinkley BR, Boughter JM (1975). Science 187:948.
Maul GG (1977). Int Rev Cytol Suppl 6:75.
Maul GG (1980). Exp Cell Res 129:431.
Schumm DE, Webb TE (1974). Biochem Biophys Res Commun 58: 354.
Steward SE, Clawson GA, Rottman FM, Patterson RY (1977). J Cell Biol 72:57.

THE MAJOR KARYOSKELETAL PROTEINS OF OOCYTES AND ERYTHROCYTES OF XENOPUS LAEVIS

Georg Krohne[1], Reimer Stick[2], Peter Hausen[2], Jurgen A. Kleinschmidt[1], Marie-Christine Dabauvalle[1], Werner W. Franke[1]

[1] Division of Membrane Biology and Biochemistry Institute of Cell and Tumor Biology, German Cancer Center, D-6900 Heidelberg, and [2] Abteilung für Zellbiologie, Max-Planck-Institut für Virusforschung, D-7400 Tubingen, Federal Republic of Germany

The nucleus ("germinal vesicle") of the full-grown vitellogenic oocyte of Xenopus laevis has an enormous size (ca. 0.4-mm diameter). It can be manually isolated and cleaned under a stereomicroscope, and with some practice, it is possible to subfractionate this nucleus into a pure nucleus envelope (NE) and the gelled nuclear content containing the amplified nucleoli, the meiotic chromosomes and nucleoplasmic particles (for details see Scheer 1972; Krohne et al. 1978a). This cell nucleus is therefore ideal to clarify the question which karyoskeletal proteins, i.e. proteins insoluble in solutions containing high salt concentrations and non-ionic detergents, are located in the nuclear interior and in the NE (for discussion of this problem in somatic cells of mammals see Berezney 1979; Franke et al. 1980; Kaufmann et al. 1981).

RESULTS

When whole nuclei of Xenopus oocytes were treated with nucleases followed by an incubation in buffer (10 mM Tris-HCl, pH 7.2) containing 1.0 M KCl and 1% Triton X-100 only two major polypeptides with apparent molecular weights (M_r) of 145,000 and 68,000 are detectable, by sodium dodecyl sulfate-polyacrylamide gel electrophoresis (SDS-PAGE)

(Laemmli 1970), in the residual karyoskeletal fraction (Fig. 1b). Both polypeptides also differ by isoelectric focusing (M_r 145,000: pH value ca. 6.15; M_r 68,000: pH values of isoelectric variants 6.4-6.6; Franke et al. 1981; Krohne et al. 1981) and tryptic peptide analysis (Krohne et al. 1982; c.f. Elder et al. 1977).

To clarify whether these polypeptides are architectural components of the NE or components of the nuclear interior isolated NE and nuclear contents were extracted with high-salt buffer and Triton as described above. In the residual NE fraction only M_r 68,000 polypeptide was highly enriched (Fig. 1c) whereas the M_r 145,000 polypeptide was only detectable in the residues of extracted nuclear contents (Fig. 1d).

Electron microscopic investigation of NE residues obtained after extraction with 1.0 M KCl and 1% Triton X-100 showed that the nuclear pore complexes and the pore-complexes-connecting fibrils (Scheer et al. 1976), equivalent to the nuclear lamina material described in other cells (Aaronson and Blobel 1975; Gerace et al. 1978), were the only structural components of the NE which were resistant to this extraction (Krohne et al. 1978a; 1981). When nuclear contents, after treatment with nucleases, were further extracted in high-salt buffer as described above residual structures of the amplified nucleoli were still visible by light and electron microscopy (Franke et al. 1981). This residual nucleolar material appeared as a meshwork of ca. 4-nm filaments which are often coiled into thicker fibrils, and form local aggregates, especially in the nucleolar cortex. The analysis of mass-isolated amplified nucleoli (Franke et al. 1981) showed that the M_r 145,000 polypeptide was highly enriched in this fraction.

For a structural and biochemical comparison with the oocyte nucleus, erythrocyte nuclei were extracted with nucleases, high-salt concentrations and non-ionic detergents as described above. The electron microscopic examination of the final fraction showed that here the nuclear lamina was the only identifiable structure in the karyoskeletal residues (Krohne et al. 1981). The chromatin and the other material of the nuclear interior was completely extracted.

Two major polypeptides with M_r 72,000 (L_I) and M_r 68,000 (L_{II}) were predominant in the karyoskeletal residues of

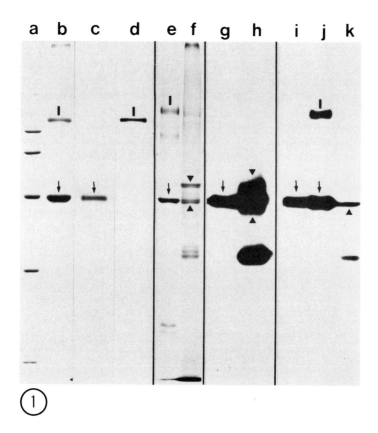

Fig. 1. SDS-PAGE (10% gel, a-d; 12% gel, e-f) of polypeptides derived from nuclear fractions of oocytes (b-e,g,i, j) and erythrocytes (f,h,k) from Xenopus laevis seen after staining with Coomassie blue (b,e,f) and as autoradiofluorograms (c,d) of proteins [^{3}H]-dansylated in vitro (Krohne et al. 1978a). Autoradiographs of the same fractions after immunoblotting with serum I (g,h) and serum II (i-k) are shown.

a: Reference proteins (Coomassie blue) are, from top to bottom, β-galactosidase (M_r 135,000), phosphorylase-a (M_r 94,000), bovine serum albumin (M_r 68,000), α-actin from rabbit skeletal muscle (M_r 42,000) and chymotrypsinogen (M_r 25,000);

b,e,g,j: karyoskeletal material from oocytes obtained after extraction of mass-isolated oocyte nuclei (Scalenghe et al. 1978) with 1.0 M KCl and 1% Triton X-100;

c: residue of 40 manually isolated NE after extraction with 1.0 M KCl and 1% Triton X-100;

d: residue of 40 manually isolated nuclear contents after extraction (see c);

f,h,k: nuclear pore complex-lamina fraction of erythrocytes;

i: 250 manually isolated whole NE. The M_r 145,000 polypeptide is marked by vertical bars, the M_r 68,000 polypeptide by arrows, and the major polypeptides of the pore complex-lamina fraction from erythrocytes by arrowheads.

Xenopus erythrocytes (Fig. 1f). Both polypeptides were different by isoelectric focusing from each other as well as from the M_r 68,000 polypeptide of the oocyte NE (pH values: L_I, 5.35; L_{II}, 6.2-6.35; cf. Krohne et al. 1981). Tryptic peptide analysis showed that the erythrocyte polypeptide L_I was totally different from the M_r 68,000 polypeptide of oocytes and that the M_r 68,000 polypeptides of the oocyte and the erythrocyte had similarities but could be clearly distinguished.

Light and electron microscopic localization was studied using two different antisera containing antibodies against the M_r 68,000 and the M_r 145,000 polypeptide of Xenopus laevis. Antiserum I was raised against the two major polypeptides (lamin A and lamin B) present in the pore complex-lamina fraction of chicken erythrocytes (Stick, Hausen 1980; cf. Shelton et al. 1980). Antiserum II was produced by immunization of a mouse with the total karyoskeletal residue of extracted Xenopus oocyte nuclei as shown in Figure 1b. Antibody reactions on gel electrophoretically separated polypeptides after the transfer to nitrocellulose paper (for details of the protein blotting see Risau et al. 1981) showed that both antisera contained antibodies recognizing the M_r 68,000 polypeptides present in the NE of oocytes and erythrocytes (Fig. 1g-k). In addition, serum I contained antibodies that reacted with the lamina polypeptide L_I of Xenopus erythrocytes (Fig. 1h). Serum II, on the other hand, had in addition antibodies that recognized the M_r 145,000 polypeptide characteristic for the interior of the oocyte nucleus (Fig. 1j). A possible proteolytic degradation product of the erythrocyte lamina polypeptides with a molecular weight of about 45,000 also showed specific reaction with the antibodies (Fig. 1h,k). The protein blotting experiments further demonstrated that in the NE oocyte only one pore complex-lamina protein, the M_r 68,000 polypeptide, was present and that polypeptide L_I which was typical for erythrocytes and other somatic cells (Krohne et al. 1981) was not detected in two-dimensional gels, either by protein staining or by immuno-blotting procedures (Stick, Krohne, 1982).

Moreover, antibody binding on proteins blotted after separation by two-dimensional gel electrophoresis showed that all isoelectric variants of the M_r 68,000 polypeptide were recognized by the specific antibodies (Stick, Krohne 1982).

Light microscopic localization using indirect immunofluorescence microscopy on frozen sections through Xenopus ovaries confirmed the distribution of the M_r 68,000 and M_r 145,000 polypeptides within the oocyte nucleus. Antibodies of serum I reacted exclusively with the NE of oocytes and somatic cells (Fig. 2a) but did not stain structures of the nuclear interior. Antibodies present in serum II also reacted with the oocyte NE but in addition decorated the amplified nucleoli and small nucleolus-like bodies abundant in these nuclei (Fig. 2b). The localization experiments with serum II clearly demonstrated that the karyoskeletal protein of M_r 145,000 characteristic for the interior of the oocyte nucleus was associated with the amplified nucleoli. These data were in agreement with results of fractionation studies (Franke et al. 1981).

In order to clarify whether the M_r 68,000 polypeptide of the oocyte NE is located in the nuclear lamina or in the pore complexes or in both structures, electron microscopic immunolocalization experiments were performed on manually isolated, purified and unfixed Xenopus oocytes, using peroxidase-conjugated antibodies (for details see Graham, Karnovsky 1966; Krohne et al. 1978b; Stick, Krohne 1982). Antibodies present in serum I decorated only the nucleoplasmic surface of the NE (Fig. 3a) and no significant reaction was found on the outer nuclear membrane and the pore complexes (Fig. 3b) when compared with NE after incubation with the pre-immune serum (Fig. 3c,d). When NE were first treated with buffered

0.5% Triton X-100 for removal of the inner and outer nuclear membrane and then incubated with serum I again, the nuclear lamina was intensely stained, with no significant reaction in the pore complexes (data not shown).

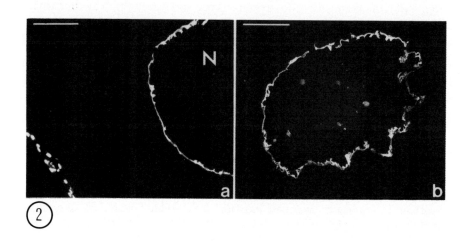

Fig. 2. Indirect immunofluorescence microscopy on frozen sections through Xenopus ovary showing oocyte nuclei with (a) and without (b) surrounding epithelial cells seen after reaction with antiserum I (a) and antiserum II (b). Note the staining of the amplified nucleoli in b. N, nucleus. Bars denote 100 μm (a,b).

Antibodies present in serum II also strongly stained the nucleoplasmic side of the NE (Fig. 3e,i) but, in contrast to serum I, these antibodies also showed strong reaction with the whole pore complex. This was obvious in both cross-sections and grazing sections (Fig. 3e,f). Frequently, the whole pore complex channel seemed to be densely covered with DAB-reaction product. In an additional experiment, isolated NE were processed for electron microscopy after incubation

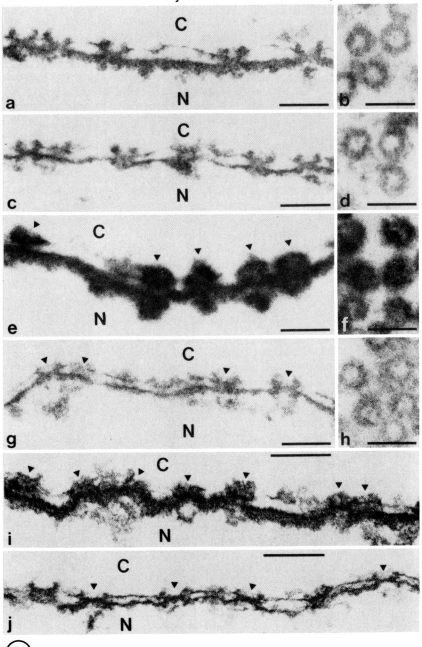

Fig. 3. Electron micrographs of ultra-thin cross-sections and tangential sections through NE of Xenopus oocytes after reactions with antibodies present in serum I (a,b), serum II (e,f,i) and the corresponding pre-immune sera (I:c,d; II:g,h,j). The bound antibodies were visualized by the peroxidase method (a-h). In i and j the DAB-reaction was omitted and ultra-thin sections were stained with uranylacetate and lead-citrate. N, nucleoplasmic side; C, cytoplasmic side. Arrowheads indicate pore complexes (e,g,i,j). Bars denote 0.2 μm (a-j).

with serum II and peroxidase-conjugated antibodies but without DAB-reaction (Fig. 3i,j). The antibodies bound were now directly visible as a fuzzy layer at the inner nuclear membrane and as a fuzzy coat on the cytoplasmic and the nucleoplasmic annuli of the pore complexes (Fig. 3i). No such additional material was seen in the corresponding control (Fig. 3j).

DISCUSSION

Our results show that the oocyte nucleus of Xenopus laevis contains two major architectural proteins (M_r 145,000 and M_r 68,000 polypeptide) which are different not only by their biochemical properties but also by their distribution within the oocyte nucleus. The M_r 145,000 polypeptide is exclusive to the nuclear interior and is associated with the amplified nucleoli and the small nucleolus-like bodies (Franke et al. 1981; Krohne et al. 1982) whereas the M_r 68,000 polypeptide is only present in the nuclear lamina and the pore complexes (Krohne et al. 1978a, 1981; Stick, Krohne 1982).

In somatic cells of mammals and birds, up to now only skeletal proteins of the pore complex-lamina have been localized (Gerace et al. 1978, Ely et al. 1978, Krohne et al. 1978b, Stick, Hausen 1980) but no architectural protein has been identified in the nuclear interior, especially in the nucleolus.

Localization experiments with serum I and serum II (Fig. 3) have shown that a certain proportion of the M_r 68,000 polypeptide is located in the nuclear lamina and another part of the M_r 68,000 polypeptide is present in the pore complex. The present data, however, do not allow a decision whether these localization experiments reflect the presence of only one polypeptide of M_r 68,000 (with several isoelectric variants) in the lamina and in the pore complex or whether both structures contain two different skeletal proteins with identical mobility in SDS-PAGE. Further protein-chemical work is necessary to characterize the isoelectrically different components of the oocyte NE protein and to clarify how close the NE protein(s) of the oocyte is related to the M_r 68,000 polypeptide (L_I) of the erythrocyte nucleus.

Aaronson RP, Blobel G (1975). Isolation of nuclear pore complexes in association with a lamina. Proc Natl Acad Sci USA 72:1007.
Berezney R (1979). Dynamic properties of the nuclear matrix. In Busch H (ed): "The Cell Nucleus," Vol 7, New York: Academic Press, p. 413.
Elder, JH, Pickett II RA, Hampton J, Lerner RA (1977). Radioiodination of proteins in single polyacrylamide gel slices. J Biol Chem 252:6510.
Ely, S, D'Arcy A, Jost E (1978). Interaction of antibodies against nuclear envelope-associated proteins from rat liver nuclei with rodent and human cells. Exp Cell Res 116:325.
Franke, WW, Scheer U, Zentgraf H, Trendelenburg MF, Muller U, Krohne G, Spring H (1980). Organization of transcribed and nontranscribed chromatin. In McKinnell RG et al. (ed): "Results and Problems in Cell Differentiation," Vol 11, Berlin, Heidelberg, New York: Springer-Verlag, p 15.
Franke WW, Kleinschmidt JA, Spring H, Krohne G, Grund C, Trendelenburg MF, Stoehr M, Scheer U (1981). A nucleolar skeleton of protein filaments demonstrated in amplified nucleoli of Xenopus laevis. J Cell Biol 90:289.
Gerace L, Blum A, Blobel G (1978). Immunocytochemical localization of the major polypeptides of the nuclear pore complex-lamina fraction. J Cell Biol 79:546.
Graham RC, Karnovsky MJ (1966). The early stages of adsorption of injected horseradish peroxidase in the proximal tubules of mouse kidney: ultrastructural cytochemistry by a new technique. J Histochem Cytochem 14:291.
Kaufmann SH, Coffey DS, Shaper JH (1981). Consideration in the isolation of rat liver nuclear matric, nuclear envelope, and pore complex lamina. Exp Cell Res 132:105.

Krohne G, Franke WW, Scheer U (1978a). The major polypeptides of the nuclear pore complex. Exp Cell Res 116:85.

Krohne G, Franke WW, Ely S, D'Arcy A, Jost E (1978b). Localization of a nuclear envelope-associated protein by indirect immunofluorescence microscopy using antibodies against a major polypeptide from rat liver fractions enriched in nuclear envelope-associated material. Cytobiology 18:22.

Krohne G, Dabauvalle MC, Franke WW (1981). Cell type-specific differences in protein composition of nuclear pore complex-lamina structures in oocytes and erythrocytes of Xenopus laevis. J Mol Biol 151:121.

Krohne G, Stick R, Kleinschmidt JA, Moll R, Franke WW, Hausen P (1982). Immunological localization of a major karyoskeletal protein in nucleoli of Xenopus oocytes. J Cell Biol, submitted.

Laemmli UK (1970). Cleavage of structural proteins during the assembly of the head of bacteriophage T4. Nature 227:680.

Risau W, Saumweber H, Symmons P (1981). Monoclonal antibodies against a nuclear membrane protein of Drosophila. Exp Cell Res 133:47.

Scalenghe F, Buscaglia M, Steinheil C, Crippa M (1978). Large scale isolation of nuclei and nucleoli from vitellogenic oocytes of Xenopus laevis Chromosoma (Berl.) 66:299.

Scheer U (1972). The ultrastructure of the nuclear envelope of amphibian oocytes. IV. On the chemical nature of the pore complex material. Z Zellforsch 127:127.

Scheer U, Kartenbeck J, Trendelenburg MF, Stadler J, Franke WW (1976). Experimental disintegration of the nuclear envelope. J Cell Biol 69:465.

Shelton KR, Higgins LL, Cochran DL, Ruffolo Jr JJ, Egle PM (1980). Nuclear lamins of erythrocyte and liver. J Biol Chem 255:10978.

Stick R, Hausen P (1980). Immunological analysis of nuclear lamina proteins. Chromosoma (Berl.) 80:219.

Stick R, Krohne G (1982). Immunological localization of the major architectural protein associated with the nuclear envelope of Xenopus laevis oocyte. Exp Cell Res, submitted.

CHARACTERIZATION OF AN ISOLATED NUCLEAR MATRIX FRACTION OBTAINED FROM EMBRYOS OF DROSOPHILA MELANOGASTER

Paul A. Fisher, Miguel Berrios and Gunter Blobel

The Laboratory of Cell Biology
The Rockefeller University
1230 York Avenue
New York, New York 10021

In recent years, considerable attention has been focused on the characterization of the various structural elements found in eukaryotic cells. Included in these studies have been a number of investigations dealing with the cell nucleus. As a result, it is now possible to conceptually subdivide the nucleus into two, apparently distinct structural domains. One, the peripheral domain, appears to be composed of the nuclear lamina and associated pore complexes, while the other, the internal nuclear matrix, constitutes a poorly defined meshwork of fibrous material traversing the nuclear interior.

Characterization of the nuclear lamina and associated pore complexes has been performed primarily with vertebrate tissues and may be summarized briefly as follows (see primary reports by Aaronson and Blobel 1974, 1975; Dwyer, Blobel 1976; Gerace et al. 1978; Gerace, Blobel 1980, 1982; Shelton et al. 1976, 1980; Lam, Kasper 1979a, 1979b; Krohne et al. 1978a, 1978b; Jackson 1976; Shelton, Egle 1979). A subnuclear fraction, apparently composed exclusively of peripheral lamina and pore complexes has been isolated. Sodium dodecyl sulfate (SDS) gel analyses have shown these fractions to contain three major polypeptides with molecular weights between 60 and 80 kilodaltons (K); these have been designated lamins A, B and C. In addition, a major high molecular weight polypeptide (150-200K) has also been observed and shown by periodic acid-schift (PAS) staining to be a glycoprotein (Bornens, Kasper 1973; Gerace, Blobel, unpublished observations). Immunocytochemical studies of the three lamins have shown them to localize exclusively in the nuclear periphery but not in the nuclear pore complexes

(Gerace et al. 1978; Krohne et al. 1978a) and to reversibly disassemble into monomeric components during mitosis (Gerace et al. 1978; Gerace, Blobel 1980).

Using preparative procedures similar to those employed in the isolation of nuclear pore complex-lamina fractions, several workers have been able to demonstrate, in addition to peripheral structures, a meshwork located internally and designated the nuclear matrix (see primary reports by Berezney, Coffey 1974, 1977; Faiferman, Pogo 1975; Comings, Okada 1976; Herlan, Wunderlich 1976; Hodge et al. 1977; Wunderlich, Herlan 1977; Herman et al. 1978; Miller et al. 1978; Herlan et al. 1979; Mitchelson et al. 1979; Long et al. 1979; van Eekelen, van Venrooij 1981). Despite the fact that by SDS-polyacrylamide gel electrophoresis (SDS-PAGE), nuclear matrix fractions appear quite similar to pore complex-lamina preparations, the results of the immunocytochemical studies cited above (Gerace et al. 1978; Krohne et al. 1978a) would seem to exclude the lamin polypeptides from the nuclear interior. At present there are no positive data clearly associating any polypeptide or group of polypeptides with the internal matrix structure; nevertheless, it seems equally clear that the structure is proteinaceous in nature and that nucleic acid is not required for its integrity.

Past studies of nuclear structure in mammalian tissues have been limited at the outset by the difficulties experienced in raising rabbit antibodies to the major pore complex-lamina antigens. We therefore decided to develop a system for studying nuclear structure which made use of an organism evolutionarily distant from convenient mammalian sources of immunoglobulin. As a further consideration, our ongoing interest in the mechanisms of mitotic disassembly of the nuclear envelope precluded the use of a species which undergoes the so-called closed type of mitosis characteristic of lower eukaryotes. An animal which fulfills both these basic criteria is <u>Drosophila melanogaster</u>. Using embryos of this organism, we have been able to purify nuclei in high yield and from these nuclei, we have been able to generate a nuclear matrix fraction (Fisher et al. 1982). In the following article the structural and biochemical properties of this fraction are described as are the initial results of our immunochemical and immunocytochemical investigations of <u>Drosophila</u> nuclear structure.

RESULTS

Morphology of Drosophila Subcellular Fractions

The procedures used to isolate nuclei from Drosophila embryos simply involved repeated low speed pelleting of the nuclei from Dounce homogenates of dechorionated whole embryos (Fisher et al. 1982). The nuclei purified in this manner were shown to be free of gross cytoplasmic contamination by both phase-contrast, light, and transmission electron microscopy. When these nuclei were digested with RNase A plus DNase I followed by sequential extraction with 2% Triton X-100 and 1 M NaCl, a subnuclear fraction was generated which, by transmission electron microscopy, was similar or identical to nuclear matrix preparations isolated from a variety of vertebrate tissues. Nucleolar remnants and an internal meshwork as well as peripheral lamina and pore complexes were all readily identified.

Chemical Composition of the Drosophila Nuclear Matrix

When the Drosophila nuclear matrix fraction was analyzed by standard techniques, it was found by mass to be approximately 93% protein, 0-1% DNA, 1-2% RNA and 4-5% cold-acid soluble nucleotide; quantitative determinations of carbohydrate and lipid were not performed (Fisher et al. 1982). Relative to the material from which it was derived, the Drosophila nuclear matrix fraction represented approximately 2% of the total embryo protein and approximately 40% of the total nuclear protein.

SDS-PAGE of the Drosophila Nuclear Matrix

When the Drosophila nuclear matrix was examined by SDS-PAGE, an extremely complex pattern was obtained (Fisher et al. 1982). In contrast to most vertebrate matrix and pore complex-lamina preparations, there appeared to be only a single major polypeptide between 60 and 80K. This polypeptide, with an estimated M_r of 74K was almost exactly coincident with the largest of the three rat liver lamins, lamin A. (Because of the large amounts of protein seen throughout the gel lane, the presence of minor species coincident with lamins B and/or C may have been obscured.) Also distinctive were major polypeptides at 174K, 103K, 53K,

42K and 16K. Of these the 17K and 42K species were relativemost abundant by mass while the 16K polypeptide was of comparable abundance stoichiometrically.

Effect of Various Extraction Conditions on the Morphological and Biochemical Integrity of the Drosophila Nuclear Matrix

In light of the ongoing controversy regarding preparation of nuclear matrix fractions from mammalian cells, we examined the effects of varying extraction conditions on both the quantity and quality of nuclear matrix material obtained. In contrast to results obtained with mammalian systems (Kaufmann et al. 1981), rapid preparation (<4 hr) of Drosophila nuclear matrices in the presence of high concentrations of N-ethyl maleimide (NEM) and with high concentrations of RNase A included prior to DNase, Triton X-100 and 1 M NaCl led to optimal recovery of large amounts of structurally intact material. When NEM or other suitable inhibitors of sulfhydryl proteases were omitted from the initial extraction buffer or when the time of preparation was intentionally increased 2- to 3-fold, the recovery of matrix protein was reduced by 50-75% and the material which was recovered was highly fragmented. Nevertheless, even in the presence of 2-mercaptoethanol or dithiothreitol, the final matrix fractions prepared in the absence of sulfhydryl protease inhibitors could be clearly shown to contain a signficant proportion of intact nuclear matrix structures in addition to the more abundant fragmented forms.

Production and Characterization of Antibodies to Drosophila Nuclear Matrix Polypeptides

Our initial attempts to generate antibodies directed against Drosophila nuclear matrix polypeptides involved injecting 100-200 μg of the 174K, 74K and 16K species (purified by a single step of SDS-PAGE) each into a different rabbit. With all three antigens, we were able to elicit high titer antisera within 1-2 months (Fisher et al. 1982). In addition, an antiserum directed against the entire Drosophila nuclear matrix fraction was generated by injection of unfractionated matrix antigen.

After initial screening by a nitrocellulose based immunoadsorption spot assay (Fisher et al. 1982), antisera were

further characterized by probing nitrocellulose blots of one dimensional SDS-polyacrylamide gels. From these blot analyses, we were able to determine that each of the three antisera raised against the molecular weight selected matrix antigens recognized only one polypeptide or group of polypeptides coinciding in size to the antigen injected. In contrast, the antiserum raised against the entire Drosophila matrix fraction recognized virtually every polypeptide identifiable on SDS-polyacrylamide gels.

Immunochemical Characterization of the Drosophila Nuclear Matrix Fraction and Comparison with the Rat Liver Nuclear Pore Complex-Lamina

In order to assess the effectiveness of our subcellular fractionation, we prepared blots containing both the whole embryo homogenate and the first post-nuclear supernatant (presumably a cytoplasmic fraction). As expected, the Drosophila nuclear matrix antigens were easily identifiable in the whole embryo homogenate but were undetectable in comparable amounts of post-nuclear supernatant. When the matrix fraction itself was similarly examined, the enrichment of matrix polypeptides observed relative to the whole embryo homogenate was approximately 25- to 50-fold, consistent with the expectation based on quantitation of total protein in the various subcellular fractions.

Cross antigenicity between the Drosophila nuclear matrix fraction and an operationally comparable mammalian subnuclear fraction, the rat liver pore complex-lamina, was assessed using techniques of nitrocellulose blot analysis; the results may be summarized as follows. When chicken antisera specific for either rat liver lamins A and C or lamin B (generous gifts of Dr. L. Gerace) were used to probe nitrocellulose blots of the Drosophila nuclear matrix, there was no discernible labeling of any Drosophila polypeptides. Similarly, the three molecular weight specific anti-Drosophila antisera showed little or no reactivity with rat liver pore complex-lamina fractions. Somewhat in contrast to these two results, however, an antiserum raised in rabbits against the entire rat liver-pore complex-lamina fraction which could be shown by blot analysis to react with all of the identifiable polypeptides in that fraction did label to a small extent, virtually all of the Drosophila nuclear matrix polypeptides as well.

A High Molecular Weight Glycoprotein is a Common Component of Subnuclear Fractions from Drosophila, Chickens, Opossums, Rats and Guinea Pigs

The glycoprotien content of the Drosophila nuclear matrix was assessed by probing nitrocellulose gel blots of the matrix polypeptides with radioactively labeled concanavalin A (Con A) (Fisher et al. 1982). Using this technique, we were able to identify a single predominant glycoprotein with an apparent M_r of 174K, i.e. coincident with the major high molecular weight polypeptide identified by Coomassie blue staining. Two-dimensional analyses, however, clearly demonstrated that the 174K glycoprotein was distinct from the major 174K Drosophila matrix polypeptide and could be completely resolved from it. In terms of protein stain, the 174K glycoprotein represents less than 5% of the total 174K protein. When subnuclear fractions from chickens, opossums, rats and guinea pigs were examined by Con A probing of nitrocellulose blots, a major polypeptide coincident with the 174K Drosophila nuclear matrix glycoprotein was specifically labeled.

In Situ Localization of the 74K Drosophila Nuclear Matrix Polypeptide by Indirect Immunofluorescence

A 1:200 dilution of our high titer anti-74K antiserum was used to stain several different tissues obtained both from Drosophila third instar larvae and from a developmentally heterogeneous mixture of Drosophila embryos (Fisher et al. 1982). For all cell and tissue types examined, there was an intense and qualitatively similar pattern of nuclear fluorescence observed with little or no cytoplasmic staining. Further, although a relatively more intensely stained nuclear rim was demonstrable, there also appeared to be areas of increased fluorescence intranuclearly.

DISCUSSION

The results presented in this article describe the purification of nuclei from Drosophila melanogaster embryos and their subfractionation to produce a nuclear matrix preparation similar to that obtained from a variety of eukaryotic tissues. To date, there are only a few reports detailing the preparation of a nuclear matrix from invertebrate sources (Herlan, Wunderlich 1976; Wunderlich, Herlan 1977; Herlan et al. 1979; Mitchelson et al. 1979) and these have all involved lower eukaryotes. Thus as far as we are aware, our work on Drosophila (Fisher et al. 1982) constitutes the first evidence of an isolable nuclear matrix in a higher eukaryotic invertebrate.

Drosophila melanogaster provides several advantages as a system in which to study nuclear structure and function. From a technical standpoint, it has proven extremely simple to prepare large quantities of nuclear matrix protein in a very short period of time. Further, the matrix structures so obtained are remarkably stable and can withstand a variety of mechanical stresses. Such stability is apparently a distinctive feature in comparison to mammalian matrix preparations and extends to certain chemical manipulations as well. Thus, in contrast to the recent report on the rat liver nuclear matrix (Kaufmann et al. 1981), routine fractionation of the Drosophila matrix in the presence of NEM and RNase A (at the appropriate stages of the purification) is without ill effect on the integrity of the isolated matrix. In fact, what does seem clear is that the Drosophila matrix is extremely sensitive to proteases and specifically, that the inhibition of sulfhydryl proteases is essential for optimal matrix preparation.

Another of the more immediate advantages of working with Drosophila is the relatively high antigenicity of the nuclear matrix proteins. High titer rabbit antisera to polypeptides throughout the common molecular weight range (16-174K) have been easily obtained and have proven extremely useful in our characterization of the Drosophila nuclear matrix. Thus, we have been able to demonstrate that matrix antigens are apparently confined to the nuclear pellet fraction and are absent from the post-nuclear supernatant. Further, the fact that the molecular weight specific antibodies we have generated all appear to react with only a single polypeptide when analyzed on one-dimensional gel blots argues strongly against the occurrence of significant nonspecific proteolytic degradation during the course of our purification.

The nuclear matrix fraction from Drosophila embryos appears to contain only a single major polypeptide (74K) in the region of the vertebrate lamins. Further, the Drosophila matrix contains substantially greater amounts of a high molecular weight (174K) polypeptide than comparable vertebrate preparations. At the present time, the significance of these observations is obscure, particularly since we have been unable to observe specific crossreactivity between either anti-Drosophila matrix antisera and rat liver pore complex-lamina polypeptides, or between chicken anti-rat liver lamin antisera and the Drosophila matrix. However, the broad crossreactivity observed with a rabbit anti-rat liver pore complex-lamina antiserum and the Drosophila matrix fraction suggests that immunochemical homologies do in fact exist. Thus, by using techniques of immunoaffinity chromatography, the specific nature of these homologies should be elucidated.

Through the technique of Con A affinity labeling, we have been able to demonstrate the presence of a high molecular weight glycoprotein in the Drosophila nuclear matrix as well as in subnuclear fractions from a variety of vertebrate tissues. Although this glycoprotein has a mobility in one-dimensional gels exactly coincident with the major 174K matrix polypeptide, two-dimensional analyses demonstrate a clear separation between the major Coomassie blue staining polypeptide and the glycoprotein. Similar results have recently been obtained with the rat liver pore complex-lamina fraction and thus it appears that both high molecular weight Drosophila polypeptides may have rat liver homologs. As above, immunochemical studies will be crucial to a further understanding of these observations.

Ultimately, crucial questions regarding the in vivo significance of the operationally defined nuclear matrix described above may be approached most directly by immunocytochemical localization studies. Thus our initial attempts to localize the 74K antigen by indirect immunofluorescence are most encouraging. From the results of our analyses, it seems that the 74K antigen localizes both peripherally as well as intranuclearly. Staining of sections through nuclei will be required in order to confirm this impression and in the event that such studies are confirmatory, immunoadsorption of antisera to two-dimensionally purified 74K polypeptides will be necessary to establish the precise molecular components involved. Such studies will form the basis for substantial future work on Drosophila nuclear structure.

ACKNOWLEDGMENTS

These studies were supported by Research Grant GM 27155 from the National Institutes of Health. P.A.F. is a postdoctoral fellow of the Helen Hay Whitney Foundation.

Aaronson RP, Blobel G (1974). On the attachment of the nuclear pore complex. J Cell Biol 62:746-754.

Aaronson RP, Blobel G (1975). Isolation of nuclear pore complexes in association with a lamina. Proc Natl Acad Sci USA 72:1007-1011.

Berezney R, Coffey DS (1974). Identification of a nuclear protein matrix. Biochem Biophys Res Commun 60:1410-1417.

Berezney R, Coffey DS (1977). Nuclear matrix. Isolation and characterization of a framework structure from rat liver nuclei. J Cell Biol 73:616-637.

Bornens M, Kasper CB (1973). Fractionation and partial characterization of proteins of the bileaflet nuclear membrane from rat liver. J Biol Chem 248:571-579.

Comings DE, Okada TA (1976). Nuclear proteins III. The fibrillar nature of the nuclear matrix. Exp Cell Res 103: 341-360.

Dwyer N, Blobel G (1976). A modified procedure for isolation of a pore complex-lamina fraction from rat liver nuclei. J Cell Biol 70:581-591.

Faiferman I, Pogo AO (1975). Isolation of a nuclear ribonucleoprotein network that contains heterogenous RNA and is bound to the nuclear envelope. Biochemistry 14:3808-3816.

Fisher PA, Berrios M, Blobel G (1982). Isolation and characterization of a proteinaceous subnuclear fraction composed of nuclear matrix, peripheral lamina and nuclear pore complexes from embryos of Drosophila melanogaster. J Cell Biol, in press.

Gerace L, Blum A, Blobel G (1978). Immunocytochemical localization of the major polypeptides of the nuclear pore complex-lamina fraction. Interphase and mitotic distribution. J Cell Biol 79:546-566.

Gerace L, Blobel G (1980). The nuclear envelope lamina is reversibly depolymerized during mitosis. Cell 19:277-287.

Gerace L, Blobel G (1982). The nuclear lamina and the structural organization of the nuclear envelope. Cold Spring Harbor Symp Quant Biol, in press.

Herlan G, Wunderlich F (1976). Isolation of a nuclear protein matrix from Tetrahymena macronuclei. Cytobiologie 13: 291-296.

Herlan G, Eckert WA, Kafferberger W, Wunderlich F (1979). Isolation and characterization of an RNA-containing nuclear matrix from Tetrahymena macronuclei. Biochemistry 18:1782-1787.

Herman R, Weymouth L, Penman S (1978). Heterogenous nuclear RNA-protein fibers in chromatin depleted nuclei. J Cell Biol 78:663-674.

Hodge LP, Mancini P, Davis FM, Heywood P (1977). Nuclear matrix of HeLa S3 cells. J Cell Biol 72:194-208.

Jackson RC (1976). Polypeptides of the nuclear envelope. Biochemistry 15:5641-5471.

Kaufmann SH, Coffey DS, Shaper JH (1981). Considerations in the isolation of rat liver nuclear matrix, nuclear envelope, and pore complex lamina. Exp Cell Res 132:105-124.

Krohne G, Franke WW, Ely S, D'Arcy A, Jost E (1978a). Localization of a nuclear envelope-associated protein by indirect immunofluorescence microscopy using antibodies against a major polypeptide from rat liver fractions enriched in nuclear envelope associated material. Cytobiologie 18:22-38.

Krohne G, Franke WW, Scheer U (1978b). The major polypeptides of the nuclear pore complex. Exp Cell Res 116:85-102.

Lam KS, Kasper CB (1979a). Selective phosphorylation of a nuclear envelope polypeptide by an endogenous protein kinase. Biochemistry 18:307-311.

Lam KS, Kasper CB (1979b). Electrophoretic analysis of three major nuclear envelope polypeptides. Topological relationship and sequence homology. J Biol Chem 254:11713-11720.

Long BH, Huang C-Y, Pogo AO (1979). Isolation and characterization of the nuclear matrix in Friend erythroleukemia cells: chromatin and hnRNA interactions with the nuclear matrix. Cell 18:1079-1090.

Miller TE, Huang C-Y, Pogo AO (1978). Rat liver nuclear skeleton and ribonucleoprotein complexes containing hnRNA. J Cell Biol 76:675-691.

Mitchelson KR, Bekers AGM, Wanka F (1979). Isolation of a residual protein structure from the nuclei of the myxomycete Physarum polycephalum. J Cell Sci 39:247-256.

Shelton KR, Cobbs C, Povlishock J, Burkat R (1976). Nuclear envelope fraction proteins: Isolation and comparison with the nuclear protein of the avian erythrocyte. Arch Biochem Biophys 174:177-186.

Shelton KR, Egle PM (1979). Analysis of the nuclear envelope polypeptides by isoelectric focusing and electrophoresis. Biochem Biophys Res Commun 90:425-430.

Shelton KR, Higgins LL, Cochran DL, Ruffolo JJ Jr, Egle PM (1980). Nuclear lamins of erythrocyte and liver. J Biol Chem 255:10978-10983.
van Eekelen CAG, van Venrooij WJ (1981). hnRNA and its attachment to a nuclear protein matrix. J Cell Biol 88: 554-563.
Wunderlich F, Herlan G (1977). A reversibly contractile nuclear matrix. J Cell Biol 73:271-278.

THE NUCLEAR LAMINS: THEIR PROPERTIES AND FUNCTIONS

Keith R. Shelton, Patsy M. Egle and
David L. Cochran

Department of Biochemistry
Medical College of Virginia
Virginia Commonwealth University
Richmond, VA 23298

In a proposal which has had a major effect on the study of the nuclear periphery, Aaronson and Blobel (1975) suggested that three proteins (now named lamins A, B and C) form a fibrous structure, continuous with the nuclear pores, at the interface between the inner nuclear membrane and the perinuclear chromatin. Our purpose in this article is to examine four aspects concerning these proteins: 1) how they are identified in the absence of functional assays, 2) where they are located in the nucleus, 3) what molecular contacts they form in the nucleus, and 4) what their functions are.

IDENTIFICATION OF THE LAMINS

The lamins were first considered as a group because of their co-isolation (Aaronson, Blobel 1975). The Dwyer and Blobel (1976) preparation from rat liver (see Fig. 1) has become a frequently used standard for comparative and other studies (Ely et al. 1978; Krohne et al. 1978a and b; Shelton et al. 1980a; Richardson, Maddy 1980; Maul, Avdalovic 1980).

The three lamins appeared to be closely related because they showed some immunological crossreactivity (Gerace et al. 1978; Ely, quoted in Krohne 1978b) and also because they yielded cleavage products of similar size (Cochran et

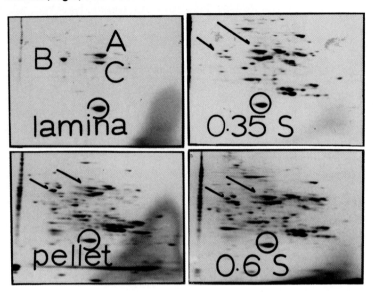

Fig. 1. Characterization of nuclear fractions by isoelectric point and molecular weight. Experimental details of electrophoresis have been described (Shelton, Egle 1979). lamina, this panel contains the proteins of the rat liver nuclear lamina fraction. Lamins A, B and C are indicated. In all panels the circled spot is carbonic anhydrase which was included as an internal standard. The other three panels contain fractions obtained by washing rat liver nuclei with solutions of increasing salt concentration (Takani, Busch 1979). Low-salt washes are not presented. Spots corresponding to the three lamins are present in each fraction although only lamins A and B are indicated by arrows. 0.35 S, protein soluble in 0.35 M NaCl. 0.60 S, protein soluble in 0.60 M NaCl. pellet, protein insoluble in these wash solutions.

al. 1979). However, further studies have shown that lamins A and B constitute two lamin types.

Resolution of the lamins by two-dimenstional gel electrophoresis, as illustrated in Figures 1 and 2, revealed that lamin B is significantly more negative and that lamin A is comprised of several isoelectric point variants (Shelton,

Nuclear Lamins / 159

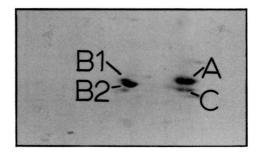

Fig. 2. Characterization of chicken erythrocyte lamins by isoelectric point and molecular weight. Lamins A, B1, B2 and C are indicated.

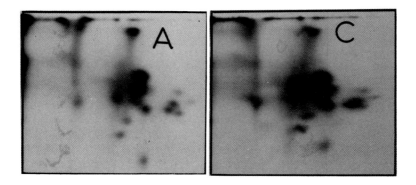

Fig. 3. Characterization of rat kidney lamins A and C by $[^{125}I]$-tryptic peptide mapping. The lamins were first purified by two-dimensional electrophoresis as shown in Figures 1 and 2. Preparation of the peptide maps has been described in detail (Shelton et al. 1980a). Quantitative comparison of these maps is not always possible because of smearing due to numerous spots and the effects of extended exposure of the x-ray film. However, visual inspection of several autoradiograms, which had been exposed to the thin-layer plates for different periods of time, indicated that these maps were virtually identical.

Egle 1979; Shelton et al. 1980a). Lamins A and B are also arranged differently in the interphase nucleus so that upon disulfide bond formation, they yield characteristically different polymerization products (Shelton, Cochran 1978; Cobbs, Shelton 1978; Lam, Kasper 1979). Finally, [^{125}I]-tryptic peptide maps of lamins previously purified by two-dimensional electrophoresis have indicated significant differences in primary sequence (Shelton et al. 1980a and b). Tryptic mapping of proteins purified by two-dimensional gel electrophoresis is recommended for the identification of lamins and their variants.

There are subtypes of each lamin type. Lamins A and C form one set of closely related subtypes. They are nearly identical by the criteria of ionic character (Shelton, Egle 1979), have a qualitatively similar topography (Cobbs, Shelton 1978; Lam, Kasper 1979), and yield very similar [^{125}I]-trypic peptide maps (Shelton et al. 1980a and b). Peptide maps of rat kidney lamins A and C are presented in Figure 3. Although these are large proteins and many spots have not been resolved, the maps are very similar. Lamins A and C would be more appropriately named A1 and A2 in order to indicate their similarity.

Although lamin B was originally thought to include a single type, there are two subtypes present in chicken cells. These have been designated lamins B1 and B2 (Fig. 2) and are very similar when compared by isoelectric focusing, [^{125}I]-tryptic peptide mapping, and _in situ_ crosslinking (Shelton et al. manuscript in preparation).

It is possible that as more cells are examined by these methods, other variations in lamin complements will be discovered. It should be noted, however, that lamin A in particular is very sensitive to chemical cleavage (Shelton 1978; Shelton et al. 1980b) and to proteolysis. For instance, protease inhibitors are unnecessary when the lamina fraction is isolated from rat liver. However, without inhibitors lamin A is significantly diminished in kidney preparations (unpublished observations). _In vitro_ cleavage products and effects must be accounted for in each study.

It can be advantageous to confirm the identity of lamins and their subtypes in two-dimensional gels without performing [^{125}I]-tryptic peptide mapping. A high sensitivity staining procedure, which yields characteristically colored spots for many proteins (GELCODE™, Upjohn Diagnostics, Kalamazoo, MI), stains lamins B1 and B2 a golden yellow. Lamin A and related spots are stained brown to orange in the spot center and black at the edges, possibly due to a concentration effect for this protein. This method is not definitive but provides for rapid screening.

Oocytes contain only one lamin-like protein. The residual nuclear envelope or non-membranous pore complex of Xenopus laevis oocytes contains a protein which migrates electrophoretically with rat liver lamin B (Krohne et al. 1978a). The comparable fraction from Spisula solidissima also contains only one protein similar to the lamins (Maul, Avdalovic 1980). However, with this protein, in vitro phosphorylation and also intermolecular crosslinking indicate a similarity with lamin C. The comparison of these oocyte proteins with the lamins of somatic cells by two-dimensional electrophoresis and [^{125}I]-tryptic peptide mapping would be of great interest.

Other proteins which are similar to lamins in molecular weight occur in the nucleus (Peters, Comings 1980). As shown by the sequential washing of nuclei with solutions of increasing salt concentration (Takami, Busch 1979), these proteins can be isolated in fractions which also contain some lamins (Fig. 1). However, they are not significantly retained in the lamina fraction and there is no evidence at present for a chemical or functional relationship between these proteins and the lamins.

ASSOCIATION OF THE LAMIN TYPES AND SUBTYPES WITH SPECIFIC NUCLEAR STRUCTURES

Membranes, peripheral chromatin and nuclear pores are closely associated at the nuclear periphery and could all contribute components to the isolated nuclear envelope fraction. Thus, lamins could be major proteins in these structures or in other structures which have been recently described such as the lamina and the nuclear matrix. In the following discussion, we summarize arguments for and against particular localizations.

Nuclear Periphery

The most direct evidence for the exclusive localization of the lamins has placed them in a thin zone at the periphery of the nucleus, either in the perinuclear chromatin or between this layer and the inner nuclear membrane. Two groups have used antibodies to the lamins with immunoperoxidase staining and ultramicroscopy. Krohne et al. (1978b) located lamin B in the peripheral chromatin at the inner nuclear envelope. Gerace et al. (1978) also located the three lamins essentially at this site, where they believe the lamina occurs. Both groups reported insignificant immunostaining of nuclear pores and the nuclear interior. These experiments have not been universally accepted as proof of exclusive localization. Several authors have argued that lamins in other regions of the nucleus might have inaccessible antigenic sites. Further, because the lamins are immunologically crossreactive, a particular lamin might be located entirely in an inaccessible site and a second lamin could give a false positive reaction at the periphery (Lam, Kasper 1979; Berezney 1980; Richardson, Maddy 1980).

Evidence that lamins A and B are major components of the lamina fraction rather than of nuclear pores has been obtained from another source (Shelton et al. 1980a). The nuclear envelope of the mature avian erythrocyte contains very few pores or pore-like structures. However, lamins A and B are the most abundant proteins in this nuclear envelope fraction. Comparison of the avian erythrocyte with rat liver has heightened the impression that the lamins are concentrated in the "lamina". Although rat liver nuclei are four-fold richer in nuclear pores (Kartenbeck et al. 1971), rat lamins A and B constitute relatively less of the total nuclear envelope protein in rat as compared with chicken. In this study, the proteins were characterized by two-dimensional electrophoresis to insure that only homologous lamins were compared. The enrichment of lamins A and B in the fraction with fewer pores has indicated their localization in the "lamina" portion.

This comparative study casts doubt on the concept that the lamins necessarily form a fibrous network. It should be noted that "lamina" isolated from chicken erythrocytes differs morphologically from the fibrous network obtained from rat liver. The chicken "lamina" was a smooth homogeneous material. It is possible that the lamins are organized in conjunction with other nuclear components and thus isolated material will differ morphologically with cell type.

Nuclear Pores

Richardson and Maddy (1980) surface-labeled intact nuclei with ^{125}I in order to differentiate between pore proteins (available at the surface) and lamina proteins (unavailable). Two proteins which appeared to be lamins A and B were labeled and therefore judged to be nuclear pore components. Another interpretation of these results is possible. Firstly, it must be noted that the lamins are the most abundant nuclear envelope proteins and are considerably more abundant than the other proteins labeled in this study (see Fig. 3 in Richardson, Maddy 1980). Secondly, approximately 12% of the nuclear surfaces were stripped of outer nuclear membrane. These stripped nuclei could contribute a significant background of labeled internal proteins. Thus, because of their relative abundance, the lamins could be represented among the labeled proteins without being present as pore components at the outer nuclear surface.

Internal Matrix

Numerous groups have obtained a nuclear fraction which appears to form a nuclear skeleton or matrix (see review by Berezney, Coffey 1976). This fraction is relatively rich in lamins and there has been controversy over whether the lamins are components of the nuclear interior as well as the nuclear periphery. Arguments based on recovery of proteins during isolation have been presented to both confirm (Berezney 1980) and deny (Gerace et al. 1978) this proposal. Some studies have dealt with specific lamins.

Krohne et al. (1978a) subfractionated rat liver nuclear membrane and obtained a membrane-free, fibrous pellet which contained only lamin C, a result suggesting that lamin C is a component of the internal matrix. This conclusion is tenuous, however. For instance, it could be argued that lamin C is the lamin most easily extracted from the membrane, and that once free of the membrane, the protein aggregates and precipitates. In regard to ease of extractability, it should be noted (see under LAMIN TOPOGRAPHY) that lamin C may be in more loosely organized polymeric structures in the nucleus than is lamin A.

One nuclear fractionation study indicated that only lamin B of the three rat liver lamins was a component of the nuclear membrane-lamina complex, and that lamins A and C were components of the matrix exclusively (Peters, Comings 1980). In the nuclear membrane-lamina complex, lamin B constituted 67.6% of the protein, and no other protein in the molecular weight range of 50,000 to 72,000 constituted more than 3.5% of the total. Although the assertion was made that these proteins accounted for the major proteins reported in other nuclear membrane and lamina fractions from rat liver, this is clearly an impossibility; one protein cannot be three. The proteins of well characterized nuclear envelope fractions have been described (Bornens, Kasper 1973; Shelton et al. 1976; Jackson 1976). Further, a series of studies characterizing the major proteins of both rat liver and chicken erythrocyte nuclear envelopes have shown these proteins to be identical with the lamins (Shelton, Cochran 1978; Lam, Kasper 1979; Shelton, Egle 1979; Shelton et al. 1980a; Gerace, Blobel 1980).

In contrast to the above studies, Long et al. (1979) used deoxycholate to remove the lamins from the nuclear matrix of Friend erythroleukemia cells. They concluded that they were not essential components of the internal matrix.

LAMIN TOPOGRAPHY

The arrangement of the lamins within the nuclear envelope has been probed by protein crosslinking experiments. Both lamins A and B are present in the envelope in homotypic oligomers. These oligomers have been converted into covalently linked polymers by forming disulfide bonds between intrinsic sulfhydryl groups, and the resultant polymers have been analyzed by a two-dimensional gel electrophoresis procedure (Shelton, Cochran 1978). This experiment provided the first evidence of a marked difference between lamins A and B. Lamin A yielded several polymers including some too large to enter the polyacrylamide gel. In contrast, lamin B was obtained as two polymers with the approximate molecular weights expected for a dimer. These experiments were performed with chicken erythrocytes in which lamin C is insufficiently abundant for this analysis, but a similar study in HeLa cells indicated that lamins A and C gave the same qualitative results (Cobbs, Shelton 1978). Further studies using rat liver revealed that lamin C yielded relatively less of each polymeric species than did lamin A (Lam, Kasper 1979). Lamin C may be more loosely arranged in the envelope, or alternatively it may lack some reactive sulfhydryls.

Lam and Kasper (1979) observed two lamin B polymers with molecular weights appropriate for trimers and suggested that the rat proteins differed from chicken lamins. This apparent difference, however, is a result of electrophoretic conditions (unpublished observations). More recent evidence, based on an examination of crosslinked, fragmented lamin B, indicates that the two polymers may be a tetramer and a trimer (Cochran, Shelton 1981).

FUNCTION OF THE LAMINS

Observed variations of the lamins in somatic cells have not suggested functions. Lamin C is decreased in avian cells as compared with mammalian cells, but it is decreased in avian cells as different as mature erythrocytes and embryonic liver (Shelton et al. 1980a). Similarly, lamin B1, the minor lamin B subtype, is found in both of these avian cells but not in mammalian cells. Thus, these appear to be class differences.

There have been no direct demonstrations of lamin functions. If they are structural proteins, which seems reasonable at this time, then demonstration of function will require a system for the reversible assembly of the nucleus in vitro and such a system may be very difficult to obtain.

A function of the lamins can be proposed, however, from indirect evidence. Their enrichment in the nuclear periphery indicates a function at that site. Their selective retention, in comparison with other nonhistone nuclear proteins, in the metabolically inactive avian erythrocyte nucleus suggests a structural role at the level of the nuclear infrastructure. Together with these suggestions, the depletion in lamin types observed in oocyte germinal vesicles, where the nuclear envelope is not associated with chromatin, suggests that lamins may function in uniting the chromatin and membrane phases of the nucleus. This proposal need not require that they form a continuous connection but only that their presence is required in some manner for the appropriate interactions of these very dissimilar phases of the nucleus.

CONCLUSION

The lamins are a class of nuclear proteins distinguished by their relative insolubility and their abundance in the nuclear envelope fraction. They appear to be concentrated in a thin zone at the nuclear periphery, either in the perinuclear chromatin or between this zone and the inner nuclear membrane.

Four lamins have been identified in somatic cells. They have been characterized by isoelectric point, molecular weight, topography and [^{125}I]-tryptic peptide map. There are two types of lamin and each type includes two subtypes; these are lamins A and C and lamins B1 and B2. It is suggested that A and C be renamed lamins A1 and A2. Other types or subtypes may occur. Lamin-like proteins have been found in the nuclear envelope of oocytes and probably constitute new lamin types or subtypes.

No function for the lamins has been demonstrated, but indirect evidence is consistent with a structural role at the interface between chromatin and the nuclear membrane.

ACKNOWLEDGMENTS

This work was supported by grants ES 02377 and CA 15923 from the Department of Health and Human Services.

Aaronson RP, Blobel G (1975). Isolation of nuclear pore complexes in association with a lamina. Proc Natl Acad Sci USA 72:1007.
Berezney R (1980). Fractionation of the nuclear matrix. Partial separation into matrix protein fibrils and a residual ribonucleoprotein fraction. J Cell Biol 85:641.
Berezney R, Coffey DS (1976). The nuclear protein matrix; isolation, structure and functions. Adv Enz Reg 14:63.
Bornens M, Kasper CB (1973). Fractionation and partial characterization of proteins of the bileaflet nuclear membrane from rat liver. J Biol Chem 248:571.
Cobbs CS, Shelton KR (1978). Major oligomeric structural proteins of the HeLa nucleus. Arch Biochem Biophys 189:323.
Cochran DL, Yeoman LC, Egle PM, Shelton KR (1979). Comparison of the major polypeptides of the erythrocyte nuclear envelope. J Supramol Struct 10:405.

Cochran DL, Shelton KR (1981). Nuclear envelope structure: identification of interpolypeptide contact sites in the lamins. J Cell Biol 91:59a.

Dwyer N, Blobel G (1976). A modified procedure for the isolation of a pore complex-lamina fraction from rat liver nuclei. J Cell Biol 70:581.

Ely S, D'Arcy A, Jost E (1978). Interaction of antibodies against nuclear envelope-associated proteins from rat liver nuclei with rodent and human cells. Exp Cell Res 116:325.

Gerace L, Blobel G (1980). The nuclear envelope lamina is reversibly depolymerized during mitosis. Cell 19:277.

Gerace L, Blum A, Blobel G (1978). Immunocytochemical localization of the major polypeptides of the nuclear pore complex-lamina fraction. J Cell Biol 79:546.

Jackson RC (1976). Polypeptides of the nuclear envelope. Biochemistry 15:5641.

Kartenbeck J, Zentgraf H, Scheer U, Franke WW (1971). The nuclear envelope in freeze etching. Ergeb Anat Entwicklungsgesch 45:7.

Krohne G, Franke WW, Scheer U (1978a). The major polypeptides of the nuclear pore complex. Exp Cell Res 116:85.

Krohne G, Franke WW, Ely S, D'Arcy A, Jost E (1978b). Localization of a nuclear envelope-associated protein by indirect immunofluorescence microscopy using antibodies against a major polypeptide from rat liver fractions enriched in nuclear envelope-associated material. Cytobiologie 18:22.

Lam KS, Kasper CB (1979). Electrophoretic analysis of the three major nuclear envelope polypeptides. Topological relationship and sequence homology. J Biol Chem 254:11713.

Long BH, Huang C-Y, Pogo AO (1979). Isolation and characterization of the nuclear matrix in Friend erythroleukemia cells: chromatin and hnRNA interactions with the nuclear matrix. Cell 18:1079.

Maul GG, Avdalovic N (1980). Nuclear envelope proteins from Spisula solidissima germinal vesicles. Exp Cell Res 130:229.

Peters KE, Comings DE (1980). Two-dimensional gel electrophoresis of rat liver nuclear washes, nuclear matrix, and hnRNA proteins. J Cell Biol 86:135.

Richardson JCW, Maddy AH (1980). The polypeptides of rat liver nuclear envelope. Examination of nuclear pore complex polypeptides by solid-state lactoperoxidase labeling. J Cell Sci 43:253.

Shelton KR (1978). The acid-dependent conversion of major nuclear oligomeric polypeptides into a smaller species. Biochem Biophys Res Commun 83:1333.

Shelton KR, Cobbs CS, Povlishock JT, Burkat RK (1976). Nuclear envelope fraction proteins: isolation and comparison with the nuclear protein of the avian erythrocyte. Arch Biochem Biophys 174:177.

Shelton KR, Cochran DL (1978). In vitro oxidation of intrinsic sulfhydryl groups yields polymers of the two predominant polypeptides in the nuclear envelope fraction. Biochemistry 17:1212.

Shelton KR, Egle PM (1979). Analysis of the nuclear envelope polypeptides by isoelectric focusing and electrophoresis. Biochem Biophys Res Commun 90:425.

Shelton KR, Higgins LL, Cochran DL, Ruffolo JJ, Egle PM (1980a). Nuclear lamins of erythrocyte and liver. J Biol Chem 255:10978.

Shelton KR, Guthrie HV, Cochran DL (1980b). On the variation of the major nuclear envelope (lamina) polypeptides. Biochem Biophys Res Commun 93:867.

Takami H, Busch H (1979). Two-dimensional gel electrophoretic comparison of proteins of nuclear fractions of normal liver and Novikoff hepatoma. Cancer Res 39:507.

THE FURTHER EVIDENCE FOR PHYSIOLOGICAL ASSOCIATION OF NEWLY REPLICATED DNA WITH THE NUCLEAR MATRIX

Bert Vogelstein, Barry Nelkin, Drew Pardoll, and Brett F. Hunt

Johns Hopkins University
School of Medicine
Baltimore, Maryland 21205

The immense amount of DNA that is contained in a eucaryotic nucleus must be organized in a way that insures proper expression, duplication and segregation of the genes. Over the past several years, three hierarchical levels of nuclear DNA organization have been delineated. The nucleosome is the most basic and well studied of these levels (Kornberg 1977; Felsenfeld 1978). Each nucleosome contains approximately 2×10^2 bp of DNA. The nucleosomes, in turn, appear to be arranged into ordered arrays of solenoids or superbeads each containing approximately 2×10^3 bp of DNA (Finch, Klug 1976; Worcel, Benyajati 1977; Renz et al. 1977; Stratling et al. 1978). At a higher order of organization, these supernucleosomal clusters appear to be packaged into topologically constrained loops, each containing about 10^5 bp of DNA (Cook, Brazell 1975; Ide et al. 1975; Benyajati, Worcel 1976; Pinon, Salts 1977; Igo-Kemenes, Zachau 1977; Hartwig 1978).

Since the hierarchical levels of DNA organization described above are present in every cell in a population, it is obvious that these structural levels of organization (as well as the DNA itself) must be duplicated in every cell cycle. Numerous studies have dealt with the question of how the nucleosomal organization of DNA is replicated during S phase (Seale 1975; Worcel et al. 1978; Riley, Weintraub 1979; Jackson, Chalkley 1981). Our studies have dealt with the way in which a higher level of DNA aggregation, that of supercoiled loops, affects the replication of DNA.

It is possible to directly visualize these loops in both metaphase chromosomes (Paulson, Laemmli 1977) and interphase nuclei (Vogelstein et al. 1980) after extraction of the histones. The loops appear as a halo of DNA anchored to a central scaffold or matrix. The nuclear matrix has been described as a consistent feature of the eucaryotic nucleus in a wide variety of cells (Berezney, Coffey 1974; Herlan, Wunderlich 1976; Hodge et al. 1977; Long et al. 1979; Mitchelson et al. 1979).

We and others have performed several biochemical and autoradiographic experiments examining the relationships between supercoiled loops, the nuclear matrix, and the process of DNA replication (Berezney, Coffey 1975; Dijkwel et al. 1979; Pardoll et al. 1980; McCready et al. 1980; Berezney, Buchholtz 1981a; Hunt, Vogelstein 1981). Based on these studies, we have proposed a simple model for the way in which nuclear DNA (in the form of topologically constrained loops) is replicated. In this communication, we will briefly summarize data concerning the relationship between the nuclear matrix and newly replicated DNA. The bulk of the paper will then be devoted to experiments which can discriminate between our interpretation of these data and various artifactual interpretations which could equally well, a priori, account for the data.

The rationale for the experimental plan that we have used is as follows. When detergent treated nuclei are incubated in a buffer containing 2 M NaCl, the histones and most other proteins are removed (Berezney, Coffey 1974; Cook, Brazell 1975; Benyajati, Worcel 1976; Peters, Comings 1980; Berezney, Buchholtz 1981b). After staining with ethidium bromide or other DNA binding agents, the nuclear DNA loops (after relaxation of supercoiling) are visualized as a halo surrounding a residual nuclear matrix (Vogelstein et al. 1980). When DNAase I is added to these matrix halo structures, DNA is progressively cleaved from them. Depending on the extent of DNAase I treatment, from zero to greater than 99% of the total nuclear DNA is cleaved from the matrix. The residual DNA that is recovered with the matrices after centrifugation represents those sequences which are near the bases of the DNA loops anchored to the matrix. If total nuclear DNA is uniformly labeled with [^{14}C]thymidine, and newly replicated DNA is labeled in vivo by administration of a short pulse of [^{3}H]thymidine, then one can easily determine if the DNA remaining associated with the matrix is random in terms of its relative content of the newly replicated DNA.

An example of such an experiment is shown in Figure 1. Nuclear matrices containing various percentages of the total nuclear DNA were made from these cells by varying the extent

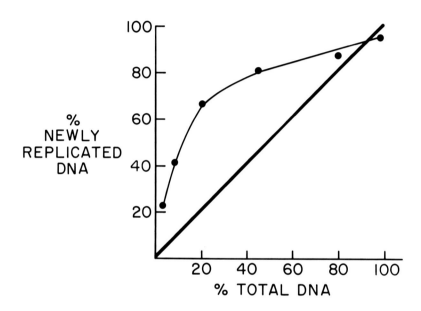

Fig. 1. Non-random association of newly replicated DNA with the nuclear matrix of mammalian cells. 3T3 cells were labeled for 48 hr with [^{14}C]thymidine and then for 30 sec with [^{3}H]thymidine. Nuclear matrices containing progressively less of the total nuclear DNA were made from these cells by varying the extent of DNAase I digestion (Nelkin et al. 1980). The percentage of total nuclear DNA ([^{14}C]thymidine radioactivity) and newly replicated DNA ([^{3}H]thymidine radioactivity), relative to that contained in the matrices in which no DNAase was used, was assessed by trichloroacetic acid precipitation of the pelleted nuclear matrices. The diagonal line represents the expected result if the distribution of newly replicated DNA with respect to the nuclear matrix were random.

of DNAase I digestion. In Figure 1, the diagonal line represents the expected result if there was a random relationship between supercoiled loops, the nuclear matrix and newly replicated DNA. As seen from Figure 1, the experimental result is decidedly non-random; 80-90% of the newly replicated DNA remains associated with the matrix when only 15-20% of the total DNA is associated.

As noted previously, nuclear DNA loops can be visualized as a halo surrounding the nuclear matrix once the histones have been removed and the supercoiling relaxed (Vogelstein et al. 1980). Autoradiography of these matrix halo structures, after administration of [^{3}H]thymidine pulses of various length, can also be used to probe the relationship between nuclear matrices, DNA loops, and newly replicated DNA. When such an experiment is performed, it is found that the newly replicated DNA (as represented by autoradiographic grains) first appears within the nuclear matrix and then gradually moves to the halo region (Vogelstein et al. 1980). In these experiments no nucleases are employed; hence, the potential non-randomness of nucleases (which could theoretically lead to artifactual results) is excluded.

Autoradiographic and biochemical experiments wherein cells were labeled with [^{3}H]thymidine for various periods of time have shown that newly replicated DNA is almost quantitatively associated with the matrix after very short pulse times, but matures into a distribution indistinguishable from the bulk nuclear DNA in 30-60 min (an interval similar to that required for replication of an individual replicon; Edenberg, Huberman 1975). We have interpreted such results as indicating that the nuclear matrix contains fixed sites for replication of DNA, that each loop of DNA is equivalent to a replicon, and that the loops of DNA are reeled through these fixed sites as they are replicated. Thus the higher order organization of nuclear DNA would be preserved directly as a byproduct of the mechanism of DNA replication (Pardoll et al. 1980; Vogelstein et al. 1980).

It should be noted, though, that newly replicated DNA is different from the bulk of nuclear DNA in several important respects. For example, newly replicated DNA contains single-stranded regions (Painter, Schaeffer 1969; Berger, Irvin 1970; Habener et al. 1970), it is arranged in chromatin in a fashion different from the bulk nuclear DNA (Seale

1975; Levy, Jakob 1978), and it is associated with proteins peculiar to the replication apparatus. These factors could potentially result in the artifactual adsorption ("sticking") of newly replicated DNA to nuclear matrix components during the extraction procedure. All the biochemical and autoradiographic results noted above showing the association of newly replicated DNA with the nuclear matrix could be explained on this basis.

To examine this alternative explanation, several experiments have been performed. Two types of "sticking" might be envisaged. In the first type, newly replicated DNA might be released from the nuclei during matrix replication, then artifactually adsorb onto other matrices. This possibility is easy to eliminate, since if it were true, one would expect most of the matrices to be labeled. In fact, autoradiographic experiments (Table 1) show that the percentage of nuclear matrices that are labeled is the same as the percentage of nuclei in S phase from which the matrices were made. Hence, if released newly replicated DNA sticks to the matrices, it must only stick to those matrices in S phase.

Table 1

Percentage of Labeled Nuclei and Nuclear Matrices from Pulse-Labeled Cells

	1-Min pulse	120-Min pulse
Nucleus	23%	26%
Nuclear matrices	21%	28%

3T3 cells were labeled with [^{3}H]thymidine for 1 min or 120 min and nuclei or nuclear matrices were prepared from these cells on coverslips as described by Vogelstein et al. (1980). The percentage of labeled nuclei or nuclear matrices was determined by autoradiography. Two hundred structures were analyzed for each measurement.

The possibility of another type of "sticking" is much harder to eliminate. It is possible that the newly replicated DNA becomes artifactually adsorbed to components within the nuclear matrix during the extraction procedures. Hence, the newly replicated DNA would never be released from those matrices from S phase nuclei, and the results of Table 1 would be obtained. Reconstitution experiments can address this possibility to some degree. As seen in Table 2, purified newly replicated DNA, and newly replicated DNA still attached to chromatin or matrix proteins, had no higher affinity for the nuclear matrices than did total nuclear DNA. However, such reconstitution experiments suffer from the fact that it is not known whether the added components can get into the matrix structures to interact with the potentially "sticky" components.

We have recently designed an experiment to directly address the possibility that single-stranded DNA adsorbs to components of the nuclear matrices during the extraction procedure. For the experiment, nuclei in which DNA was uniformly labeled with [^{3}H]thymidine, were treated with DNAase I (Riley 1980) in order to generate single-stranded regions in situ. These regions were then selectively labeled with [^{32}P]dCTP using E. coli polymerase I. When DNA was purified from nuclei treated in such a manner, it was found that the ^{32}P label was extensively associated with single-stranded regions; 60% of DNA fragments of average length 4,000 bp (generated by EcoRI digestion) contained a single-stranded region, as assessed by binding to nitrocellulose. The [^{3}H]-labeled DNA from the same nuclei was three-fold less enriched in single-stranded regions. As seen in Figure 2, however, matrices prepared from nuclei labeled in this fashion were not enriched in the [^{32}P]-labeled DNA. Comparison of Figure 2 to Figure 1 indicates that the association of newly replicated DNA with the nuclear matrices is probably not due to the artifactual "sticking" of single-stranded regions per se.

The experiment of Figure 2, however, certainly does not rule out the possibility that properties of the newly replicated DNA (other than its single-strandedness) artifactually mediate the association with the nuclear matrix. To learn more about the reality of the relationship between newly replicated DNA, supercoiled loops and the nuclear matrix, we have experimented with the acellular slime mold P. polycephalum.

Table 2

Addition of Newly Replicated DNA to Nuclei and Nuclear Matrices

	Addition to nuclei		Addition to nuclei	
	[^{3}H]-Bound (%)	[^{14}C]-Bound (%)	[^{3}H]-Bound (%)	[^{14}C]-Bound (%)
Purified DNA	1.3%	1.6%	2.3%	2.1%
Chromatin containing newly replicated DNA	0.8%	0.9%	1.1%	1.1%
Sonicated matrix containing newly replicated DNA	1.4%	1.6%	1.8%	2.0%

3T3 Cells were labeled with [^{3}H]thymidine and [^{14}C]thymidine as in Figure 1, and nuclei prepared from these cells (as in Nelkin et al. 1980) were divided into three parts. From one part, DNA was purified by SDS-protease digestion followed by phenol chloroform extraction (Gross-Bellard et al. 1973). From the second part, the nuclei were sonicated for 10 sec and then subjected to centrifugation of 15 min at 10,000 x g. The supernatant (containing greater than 95% of the ^{3}H and ^{14}C cpm) was denoted as chromatin. From the third part, nuclear matrices were prepared, then sonication and centrifugation were performed as above to yield "sonicated matrix." The purified DNA, "chromatin," or "sonicated matrix" was then added to 3T3 cell nuclei, and, after addition of an equal volume of 4 M NaCl to prepare matrices, and incubation for 10 min at 37°C (without DNAase) the matrices were subjected to centrifugation at 10,000 x g for 15 min. Similarly, additions were made to 3T3 cell matrices which had been previously treated with DNAase I (which was inactivated with EDTA before addition of the labeled DNA). The percentages of [^{3}H]- or [^{14}C]-radioactivity in the matrix pellet after centrifugation are indicated in the table.

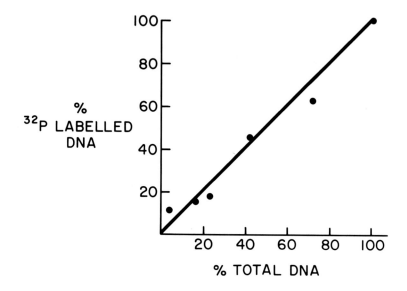

Fig. 2. Single-stranded DNA generated in nuclei in situ is randomly associated with the nuclear matrix. 3T3 cells were labeled for 48 hr with [^{3}H]thymidine. Nuclei from these cells were treated with 1 µg/ml DNAase I at 37°C for 10 min in RSB and then the single-stranded regions were preferentially labeled with [^{32}P]dCTP (Amersham, 410 Ci/mmol) using E. coli polymerase I (obtained from New England Biolabs, Beverly, MA), 5 U/ml at 15°C for 5 min. It should be noted that the amount of DNAase I used in these experiments was far in excess of that which preferentially nicks "active genes" (Levitt et al. 1979). Nuclear matrices were made from these cells and the percentage of labeled DNA remaining with the matrices was assessed as in Figure 1. The diagonal line represents the expected result if the distribution of [^{32}P]-labeled DNA with respect to the nuclear matrix were random.

According to the matrix model proposed for DNA replication, each supercoiled loop functions as a replicon. Hence it would be predicted that the length of the supercoiled loops in any nucleus should be the same as the length of the replicons in the same nucleus. Since the replicon lengths of P. polycephalum are significantly different from those of mammalian cells, an opportunity was provided for directly testing this prediction. Loop lengths were measured by fluorescence microscopy in mammalian nuclei and P. polycephalum nuclei and these lengths compared to the replicon lengths measured by DNA fiber autoradiography. The data in Table 3 shows that there is remarkable concordance between the length of the supercoiled loops and replicon lengths in nucleus from both organisms.

Table 3

Comparison of Replicon Length and Supercoiled Loop Lengths in Mammalian and Slime Mold Nuclei

Organism	Supercoiled loop length (fluorescence microscopy)	Replicon length (fiber autoradiography)
Human	75,000 bp	69,000 bp
Physarum	36,000 bp	37,500 bp

The supercoiled loop length was measured as the halo diameter of matrix-halo structures in the presence of ethidium bromide (Vogelstein et al. 1980). Loops were relaxed by a 10-sec treatment with ultraviolet illumination in the fluorescence microscope. In a variety of mammalian cells, the loops have been measured as 60-90,000 bp (B. Vogelstein, unpublished). The average replicon lengths as assessed by fiber autoradiography were obtained from published data (Hand 1977; Funderud et al. 1979).

Further insight into the relationship between DNA synthesis, supercoiled loops and the nuclear matrix can be obtained with pulse-chase experiments of the naturally synchronized nuclei in plasmodia of P. polycephalum. If a short pulse of [^{3}H]thymidine is given at the very beginning of S phase, then followed by a chase with unlabeled thymidine, then regions surrounding the origins of replication of the first replicons to be activated will be exclusively labeled. A technical problem with this experiment is to achieve an effective chase, i.e., there should be little label incorporated into DNA during the chase compared to that incorporated during the pulse. By keeping the pulse time very short (30 sec or less), we have been able to perform pulse-chase experiments with a relatively low level of label incorporation during the chase period.

When nuclear matrices are made from plasmodia which have been pulsed and chased in this way, clear-cut results are obtained (see Table 4):

(1) Immediately after the pulse, there is a significant enrichment for newly replicated DNA in the matrix fraction. This result could be due to either a physiological or artifactual association of newly replicated DNA with the nuclear matrix.

(2) After a 2-hr chase, 87% of the newly replicated DNA is released from the matrix. The relative ratio of newly replicated DNA to total nuclear DNA becomes 0.54, i.e., less than 1.0. This result could only be obtained if the replicon origins become non-randomly disassociated from the matrix during the chase period. It is consistent with the idea that replicon origins occur at the base of DNA loops and that these origins move to the periphery of the loop (farthest from the nuclear matrix anchorage points) during S phase.

The importance of the data presented in Table 4 is that it shows a non-random relationship between [^{3}H]-labeled DNA replicated at the beginning of S phase and the nuclear matrix at a time (after the chase) when this DNA is not associated with single-stranded regions, is not associated with immature nucleosomes, and is not associated with a special class of proteins. If the association of newly replicated DNA with the matrix were artifactual, and the position of replicon origins within supercoiled loop domains were random, one would expect a ratio of ^{3}H to ^{14}C DNA in the matrix fractions (after the chase) to be 1.0. The fact that this ratio is significantly less than 1.0 is strong evidence that the association of newly replicated DNA with the nuclear matrix is not an artifact of the experimental methods used.

Table 4

The Transient Association of DNA Sequences Surrounding Replicon Origins with the Nuclear Matrix

	Newly replicated DNA remaining with the matrix (^{3}H cpm)	% Newly replicated DNA on the matrix	Total DNA remaining with the matrix	% Total DNA remaining with the matrix	Enrichment
Pulse	416	37.3	317	4.8	7.8
Pulse-Chase	54	2.8	347	5.2	0.54

A macroplasmodia, previously labeled with [^{14}C]thymidine, was pulse-labeled with [^{3}H]thymidine 3 min after the second synchronous telophase. The plasmodia was then divided in half; one half was homogenized immediately and the other half was incubated in chase media containing unlabeled thymidine for 2 hr before homogenization. After homogenization, nuclei were purified and nuclear matrices were prepared from the nuclei using DNAase I digestion for 20 min. [^{3}H]- and [^{14}C]-labeled DNA in the matrix pellet were determined by acid precipitation after RNAase and protease digestion.

ACKNOWLEDGMENT

This work was supported by grants CA-06973 and CA-09243 and a gift from the Bristol-Myers Company.

Benyajati C, Worcel A (1976). Cell 9:393.
Berezney R, Buchholtz LA (1981a). Exp Cell Res 132:1.
Berezney R, Buchholtz LA (1981b). Biochemistry 20:4995.
Berezney R, Coffey DS (1974). Biochem Biophys Res Commun 60:1410.
Berezney R, Coffey DS (1975). Science 189:291.
Berger H, Irvin JL (1970). Proc Natl Acad Sci USA 65:152.
Cook PR, Brazell IA (1975). J Cell Sci 19:261.
Dijkwel P, Mullendero L, Wanka F (1979). Nuc Acids Res 6:219.
Edenberg HJ, Huberman JA (1975). Ann Rev Genetics 9:245.
Felsenfeld G (1978). Nature 271:115.
Finch, JT, Klug, A (1976). Proc Natl Acad Sci USA 73:1897.
Funderud S, Andreassen R, Haugli F (1979). Nuc Acids Res 6:1417.
Gross-Bellard M, Zudet P, Chambon P (1973). Eur J Biochem 36:32.
Habener JF, Bynum BS, Shack J (1970). J Mol Biol 49:157.
Hand R (1977). Human Genet 37:55.
Hartwig M (1978). Acta Biol med Germ 37:421.
Herlan G, Wunderlich F (1976). Cytobiologie 13:291.
Hodge LD, Mancini P, David FM, Heywood P (1977). J Cell Biol 72:194.
Hunt BF, Vogelstein B (1981). Nuc Acids Res 9:349.
Ide T, Nakane M, Anzai K, Andoh T (1975). Nature 258:445.
Igo-Kemenes WT, Zachau HG (1977). Cold Spring Harbor Symp Quant Biol 42:109.
Jackson V, Chalkley R (1981). Cell 23:121.
Kornberg RD (1977). Ann Rev Biochem 46:931.
Levitt A, Axel R, Cedar H (1979). Dev Biol 69:496.
Levy A, Jakob KM (1978). Cell 14:259.
Long BW, Huang CV, Pogo AO (1979). Cell 18:1079.
McCready SJ, Godwin J, Mason DW, Brazell IA, Cook PR (1980). J Cell Sci 46:365.
Mitchelson KR, Bekers AGM, Wanka F (1979). J Cell Sci 39:247.
Nelkin BD, Pardoll DM, Vogelstein B (1980). Nuc Acids Res 8:5623.
Painter RB, Schaeffer A (1969). Nature 221:1215.
Pardoll DM, Vogelstein B, Coffey DS (1980). Cell 19:527.

Paulson JR, Laemmli UK (1977). Cell 12:817.
Peters KE, Comings DE (1980). J Cell Biol 86:135.
Pinon R, Salts Y (1977). Proc Natl Acad Sci USA 74:2850.
Renz M, Nehls P, Hozier J (1977). Proc Natl Acad Sci USA 74:1879.
Riley DE (1980). Biochemistry 19:2977.
Riley D, Weinbraub H (1979). Proc Natl Acad Sci USA 76:328.
Seale RL (1975). Nature 255:247.
Stratling WH, Muller V, Zentgraf H (1978). Exp Cell Res 117:301.
Vogelstein B, Pardoll DM, Coffey DS (1980). Cell 22:79.
Worcel A, Benyajati C (1977). Cell 12:83.
Worcel A, Han S, Wong ML (1978). Cell 15:969.

NUCLEAR MATRIX ORGANIZATION AND DNA REPLICATION

Ronald Berezney, Joseph Basler,
Linda A. Buchholtz, Harold C. Smith,
and Alan J. Siegel

Division of Cell and Molecular Biology
Department of Biological Sciences
State University of New York
Buffalo, New York 14260

THE NUCLEAR MATRIX

Nuclei of eucaryotic cells contain a complex fibrogranular material which was earlier referred to as the interchromatinic substance (Swift 1963) but is now generally called the nuclear matrix (Berezney 1979). The nuclear matrix is the ground substance within which are immersed both chromatin and the products of chromatin transcription; the ribonucleoprotein particles. The nonchromatin nature of the in situ matrix is clearly revealed by the EDTA regressive staining procedure (Bernhard 1969). With this technique the DNA containing chromatin is either not stained or only weakly stained while the nonchromatin matrix stands out against the bleached chromatin background. The matrix appears as a fibrogranular network extending throughout the nuclear interior from the nucleolus to the nuclear pore complexes along the nuclear periphery (Fig. 1).

The reality of the in situ matrix is further substantiated by the isolation of nuclear matrix structures. The isolated matrices are the residual nuclear structures remaining after the nearly complete removal of chromatin and nuclear membrane phospholipids (Berezney, Coffey 1974). The extensive fibrogranular network in the interior of isolated matrices is strikingly similar to that of the in situ matrix (Fig. 1, Berezney, Coffey 1974, Berezney, Coffey 1977). Moreover, the isolated matrix maintains the same

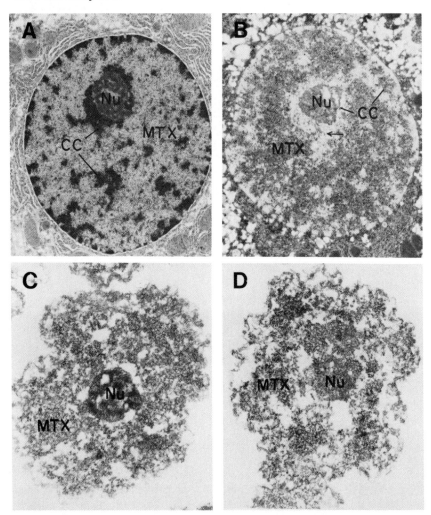

Fig. 1. Thin sectioned electron micrograph comparison of glutaraldehyde fixed, in situ rat liver nuclei and isolated nuclear matrix by standard staining with uranyl acetate and lead citrate and EDTA regressive staining (Bernhard 1969). CC, condensed chromatin, MTX, matrix; NU, nucleolus. Note the similarity in structure of the isolated matrix and in situ matrix. A, rat liver, standard staining, 9,000X; B, rat liver, EDTA regressive staining, 8,000X; C, isolated nuclear matrix, standard staining, 15,000X; D, isolated matrix, EDTA regressive staining, 13,500X.

general structural pattern wherein the internal matrix is in close apposition with the nucleolus and the surrounding pore complexes (Fig. 1).

THE CELL MATRIX

As previously suggested (Berezney, Coffey 1976), the nuclear matrix may be the nuclear extension of an overall cell matrix which is distributed throughout the cell from the surrounding plasma membrane to the nuclear interior.

A number of investigations have demonstrated a complex structure or cytoskeleton within the cytoplasm (Wolosewick, Porter 1976, Wolosewick, Porter 1979; Lenk et al. 1977; Osborn, Weber 1977). For example, Wolosewick and Porter (1976, 1979) have observed a three-dimensional microtrabecular lattice structure in the cytoplasm of whole cells. Recently we have observed an analogous three-dimensional structure in isolated nuclear matrices examined with steroscopic electron microscopy (Berezney, Siegel, manuscript in preparation). It is intriguing to consider the possible role of "structure protoplasm", be it of nuclear or cytoplasmic origin, in the organization and regulation of molecular events. Thus the overall cell matrix may represent an inherent structural order of protoplasm which is essential for both the orderly procession and complex integration of cellular processes.

STRUCTURAL DOMAINS OF NUCLEAR FUNCTION

High resolution electron microscopic autoradiographic studies indicate that both the sites of DNA replication and transcription are preferentially localized in the in situ matrix in close apposition to the condensed chromatin (Fakan, Bernhard 1971; Fakan, Hancock 1974). In addition, appropriate pulse-chase experiments have demonstrated the migration of newly transcribed extranucleolar RNA into the matrix interior (Fakan et al. 1976). Thus the in situ matrix potentially represents a functional ground substance for the processes of replication, transcription, as well as RNA processing and transport (Berezney, Coffey 1975, Berezney, Coffey 1976; Berezney, Coffey 1977; Wunderlich et al. 1976; Berezney 1979). Studies with isolated nuclear matrix support the concept of the matrix as a general functional milieu for nuclear processes (Berezney 1981). A variety of properties have been reported in association with isolated nuclear matrices including the presence of DNA attachment sites (Georgiev et al. 1978), DNA replicational sites (Berezney, Coffey 1975, Berezney, Coffey 1976; Dijkwel et al. 1979; Pardoll et al. 1980; Berezney, Buchholtz 1981a), DNA polymerase association (Smith, Berezney 1980), HnRNA interaction (Long et al. 1979; van Eekelen, van Venrooij 1981), steroid hormone and carcinogen binding sites (Barrack, Coffey 1980; Hemminki, Vainio 1979), viral protein associations (Hodge et al. 1977; Buckler-White et al. 1980), and the active phosphorylation of matrix proteins (Allen et al. 1977).

MATRIX DNA ATTACHMENT SITES

Several different experimental approaches suggest that the huge eucaryotic DNA molecules are arranged in the cell nucleus in repeating domains or loops with an average repeat size of 50 to 200 kilobases (Cook, Brazell 1975; Igo-Kemenes, Zachau 1978; Paulson, Laemmli 1977). The repeating DNA loops appear to be attached to nonhistone protein core structures, particularly the interphase nuclear matrix and its metaphase equivalent termed the scaffold (Paulson, Laemmli 1977). This suggests that the small fragments of DNA which remain tightly bound to DNA-depleted nuclear matrices (Berezney, Buchholtz 1981a), should represent the basal attachment fragments for the DNA loops. Consistent with this interpretation, our laboratory has recently isolated nuclear matrices while avoiding significant cleavage of the high molecular weight DNA (Berezney, Buchholtz 1981b). Most of the nuclear DNA remains associated with the residual matrix structures and whole mount electron microscopy has revealed that the DNA is attached to the matrix in a series of highly folded loops (Berezney, Siegel, manuscript in preparation). Furthermore, there is a linear relationship between the amount and average fragment size of the matrix-attached DNA (Fig. 2). Significantly, the average fragment size approaches zero as the amount of DNA on the matrix approaches zero.

MATRIX DNA HYBRIDIZATION AND THE DYNAMIC DNA LOOP MODEL

To determine whether specific "attachment sequences" anchor the DNA loops to the matrix, we purified small DNA

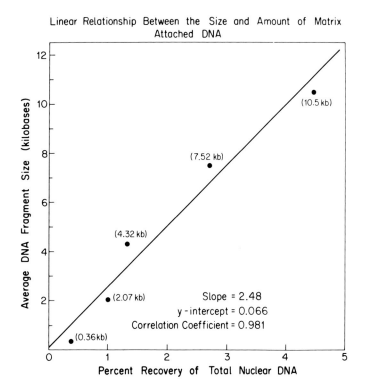

Fig. 2. Relationship between the average length and amount of DNA in isolated nuclear matrix. Matrices with different amounts of attached DNA were generated by endogenous digestion of rat liver nuclei between 10 and 120 min at 37°C. The matrix-attached DNA fragments were purified and the weighed average fragment size determined on non-denaturing 1% agarose gels with λ-DNA, PM-2-DNA and HaeIII digested PM-2 as molecular weight markers.

fragments from rat and mouse liver nuclear matrix ($\leq$350 bp) which presumably lie closest to the matrix binding sites. The reassociation kinetics of the small matrix DNA probe were virtually identical to total DNA probe in total DNA excess experiments (Basler et al. 1981). Similar results were obtained with metaphase scaffold DNA fragments (≤ 600 bp) from Chinese Hamster DON chromosomes. As shown in Figure 3, the reassociation kinetics of metaphase scaffold and interphase matrix fragments were essentially identical to total nuclear DNA probes. These findings support our results with rat and mouse liver and further demonstrate that the matrix-attached DNA is not enriched in repetitive DNA sequences in either interphase or metaphase cells.

In contradiction to the preceding results, several earlier studies have concluded that matrix and scaffold DNA are enriched in repetitive DNA sequences (Razin et al. 1978,

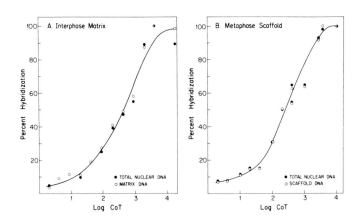

Fig. 3. Total genomic DNA-driven annealing of Chinese Hamster DON cell interphase matrix (A), and metaphase scaffold (B), DNA fragments. Matrix and scaffold DNA fragments (≤ 600 bp, O) were isolated by elution from nondenaturing agarose gels, nick translated with ^{32}P label and reassociated in the presence of an excess of total nuclear DNA driver (approximately 500 bp) and a homologous total DNA probe (approximately 400 bp, ●) nick translated with ^{3}H label. Hybridization was measured by the S_1 method. All values were normalized such that the maximum extent of hybridization of each probe represented 100%.

Razin et al. 1979; Jeppesen, Bankier 1979). We have further examined this possibility by annealing mouse liver matrix DNA ($\leq$350 bp) to a library of cloned repetitive mouse DNA sequences including mouse major satellites. Our results indicated a similar proportion of specific repetitive sequences in matrix and total nuclear DNA (Basler et al. 1981). Since a repetitive sequence repeated more than 10,000 times in the genome should be present in the clone library, these data make it highly unlikely that a specific repetitive sequence family could serve as the basis for the approximately 20,000-60,000 matrix attachment sites.

These results suggest that the DNA loops are not attached to the matrix at permanently fixed sites along the DNA molecule. To clarify this point we performed matrix DNA-driven hybridizations. If permanently fixed attachment sites exist along the eucaryotic DNA molecule, then the small DNA fragments associated with these fixed sites should represent a very small amount of the total DNA complexity ($<$1%) and should be unable to significantly drive a total nuclear DNA probe. In contrast to this prediction, our hybridization results demonstrated that small matrix DNA fragments ($\leq$350 bp) drove total DNA probes with virtually identical kinetics and extent as the homologous probe (Basler et al. 1981).

We conclude that the postulated intranuclear loop arrangement of eucaryotic DNA is not mediated by specific attachment site sequences and that the DNA adjacent to these sites is essentially random in sequence. We further propose that the DNA loops may be dynamic rather than statically fixed structures. In this manner, the number of attachment sites is preserved while the actual sites of DNA interaction are subject to change via a dynamic process DNA association-dissociation (Fig. 4).

MATRIX DNA REPLICATION SITES

Aside from an obvious role for the three-dimensional arrangement and packaging of chromatin, the organization of eucaryotic DNA into repeating loops may also order the DNA into distinct functional units. For instance, in the dynamic loop model which we propose (Fig. 4), the rapid exchange of DNA with a fixed number of attachment sites could provide a basis for the regulation of specific loop formation during

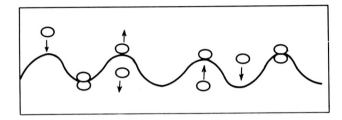

Fig. 4. Dynamic loop model of DNA. DNA attachment sites (0) are depicted in different states of association and dissociation with the DNA strand.

DNA replication and transcription. Since eucaryotic replication occurs in discontinuous replicon units with a size similar to the matrix attached loops (Berezney, Buchholtz 1981a), it is intriguing to consider whether replicons represent matrix-attached DNA loops in the process of replication.

As discussed earlier, electron microscopic autoradiographic studies have indicated a close associated of the native replicational sites with the in situ matrix structure (Fakan, Hancock 1974). Isolated nuclear matrices have enabled a direct quantitation of newly replicated DNA associated with the matrix. The earlier studies of Berezney and Coffey (1975, 1976) demonstrated an initial enrichment in matrix-associated newly replicated DNA following a 1 min in vivo pulse in regenerating rat liver. With increasing pulse time there was a rapid and progressive decrease in labeled DNA attached to the matrix and a corresponding increase in the bulk chromatin (Berezney, Coffey 1975, Berezney, Coffey 1976). These studies were the first to suggest that the DNA replicational sites are fixed to the matrix and that previously replicated DNA moves progressively away from the matrix replicational sites. Recently we repeated these experiments with improved matrix isolation techniques (Berezney, Buchholtz 1981a) and confirmed the original observations (Fig. 5).

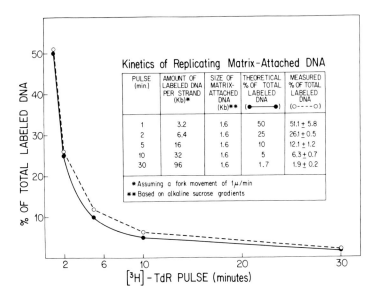

Fig. 5. Kinetics of association of replicating DNA with the nuclear matrix. Experimental data which described the kinetic association of replicating DNA attached to the matrix structure is compared with a theoretical kinetic model. In this model it is assumed that replication occurs in continous association with the matrix structure and that previously replicated DNA is found progressively further away from the "matrix replicational sites." This can be envisioned as the result of the movement of the matrix replicational sites along the replicating DNA strands or vice versa. In the theoretical calculations, the average rate of fork movement is estimated as one micron or 3.2 kilobases per minute. The average size of the 1 min labeled matrix-attached DNA fragments is estimated as 1.6 kilobases. (From Berezney, Buchholtz 1981a.)

Fig. 6 (next page). Diagrammatic representation of hypothetical structural models for matrix-bound replication. In both models the thick lines of the DNA refer to the parental strands while the thin lines refer to the newly replicated daughter strands. A: fixed matrix model. The diagram on the left indicates a region of the DNA between matrix-attached loops which is closely associated with matrix circles. Whether this corresponds to a defined "interloop

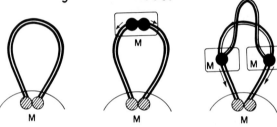

Figure 6

matrix attachment site" is unclear. In the middle diagram, a fixed bidirectional replicational complex (solid circles) becomes associated with this interloop attachment site. DNA from the two adjacent loops then move in opposite directions through the replicational complex. This forms a new looped structure containing the newly synthesized DNA (right diagram). The DNA of the newly replicated loop is therefore derived from a portion of the DNA from each of two adjacent loops. B: sliding matrix model. In the left diagram is shown a loop of unreplicated DNA attached to the matrix (hatched circles). In this model, two adjacent matrix-bound replicational complexes become attached to the DNA loop at a site other than the initial DNA attachment site (solid circles, middle diagram). Replication then proceeds in both directions by the sliding of the matrix-attached replicational complexes through the DNA in the direction of the original matrix attachment sites (right diagram). Replication of the loop terminates at these sites resulting in the formation of two copies of the original loop. (From Berezney, Buchholtz 1981a.)

Figure 6 illustrates two alternative views for replication of matrix-attached DNA loops. Although other studies of matrix-associated replication have favored the "fixed matrix" model (Dijkwel et al. 1979; Pardoll et al. 1980) we believe that more detailed information is needed concerning the interaction of replicating DNA loops with the putative matrix replicational sites before alternative views such as the "sliding matrix" model can be ruled out (see Berezney, Buchholtz 1981a).

IN VITRO DNA SYNTHESIS IN NUCLEAR MATRIX

As Bollum perceptively stated in a review article on mammalian DNA polymerase in 1975 (Bollum 1975): "... the location of a product does not describe the location of the production facility." With this in mind, our laboratory is currently studying the possible association of the DNA replicational machinery with the nuclear matrix. We have found that isolated nuclear matrices from actively replicating, regenerating liver contain significant amounts (10-25% of the total nuclear activity, Fig. 7) of DNA polymerase α activity (Smith, Berezney 1980) which is widely believed to represent the major replicative enzyme in mammalian cells. In contrast, only trace amounts of β polymerase activity (a presumptive DNA repair enzyme) are measured in the regenerating liver matrix (Fig. 7). Moreover, the matrix-bound α polymerase is completely inhibited by aphidicolin, a specific inhibitor of DNA polymerase α, but is resistant to dideoxy TTP an inhibitor of both β and γ polymerases (Smith, Berezney 1980). The absence or trace amount of α polymerase activity from slowly proliferating normal rat liver (Fig. 7) lead us to further propose that functional replicational complexes are dynamically assembled on the nuclear matrix during active DNA replication (Smith, Berezney 1980).

In support of this proposal we have recently demonstrated that the matrix-bound α polymerase effectively synthesizes DNA on the matrix-attached DNA fragments (Smith, Berezney 1981). The activity of endogenous DNA synthesis in the matrix was approximately 5.4- and 2.8-fold higher than total nuclear incorporation when expressed per mg DNA or protein, respectively. Alkaline sucrose gradients of the matrix synthesized DNA revealed one major peak at 4-5S (150-250 bp) while the bulk matrix DNA fragments sedimented

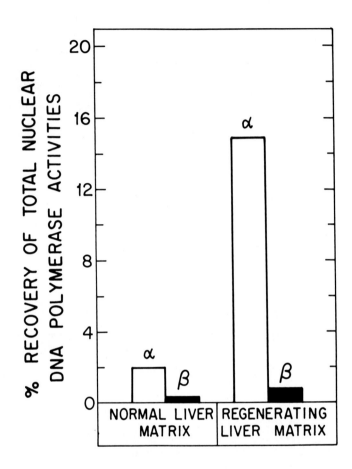

Fig. 7. Recovery of total nuclear DNA polymerase α and β activities in isolated nuclear matrices. Recoveries are based on the incorporation of ^{3}H label into acid precipitable counts on activated calf thymus DNA. It is important to note that the recovery of total α polymerase in normal liver (2%) is a maximal estimate since no α activity was detected in this fraction (see Smith, Berezney 1980).

more heterogeneously with an average of approximately 10S (1,500 bp). DNA polymerase α activity represented up to 90% of total matrix-driven DNA synthesis based on N-ethylmaleimide and aphidicolin inhibition. Moreover, the inhibition of α polymerase with N-ethylmaleimide resulted in a corresponding inhibition of the 4-5S alkaline sucrose gradient peak. Further studies of DNA synthesis in isolated nuclear matrix should be helpful in clarifying the molecular organization of the replicational complex and the molecular events of DNA replication.

ACKNOWLEDGMENTS

We wish to thank Drs. N. D. Hastie and D. Pietras for providing the clone library of repetitive mouse DNA sequences and for their valuable suggestions and help with the hybridization experiments. We are also grateful to Dr. S. Matsui for the Chinese Hamster DON cell isolated nuclei and chromosomes. Studies reported from this laboratory were supported by a USPHS research grant (GM 23922).

Allen SL, Berezney R, Coffey DS (1977). Phosphorylation of nuclear matrix proteins in isolated regenerating rat liver nuclei. Biochem Biophys Res Commun 75:111.

Barrack ER, Coffey DS (1980). The specific binding of estrogens and androgens to the nuclear matrix of sex hormone responsive tissues. J Biol Chem 255: 7265.

Basler J, Hastie ND, Pietras D, Matsui S, Sandberg AA, Berezney R (1981). Hybridization of nuclear matrix attached DNA fragments. Biochemistry, in press.

Berezney R. (1979). Dynamic properties of the nuclear matrix. Cell Nucleus 7:413.

Berezney R. (1981). The nuclear matrix: a structural milieu for the intranuclear attachment and replication of eucaryotic DNA. In Schweiger HG (ed): "International Cell Biology 1980-1981," Berlin: Springer-Verlag, p 214.

Berezney R, Buchholtz LA (1981a). Dynamic association of replicating DNA fragments with the nuclear matrix of regenerating liver. Exp Cell Res 132:1.

Berezney R, Buchholtz LA (1981b). Isolation and characterization of rat liver nuclear matrices containing high molecular weight DNA. Biochemistry 20: 4995.

Berezney R, Coffey DS (1974). Identification of a nuclear protein matrix. Biochem Biophys Res Commun 60:1410.

Berezney R, Coffey DS (1975). Nuclear protein matrix: association with newly synthesized DNA. Science 189: 291.

Berezney R, Coffey DS (1976). The nuclear protein matrix: isolation, structure and functions. Adv Enz Regul 14:63.

Berezney R, Coffey DS (1977). Nuclear matrix. Isolation and characterization of a framework structure from rat liver nuclei. J Cell Biol 73:616.

Bernhard W (1969). A new staining procedure for electron microscopical cytology. J Cell Biol 27:250.

Bollum FJ (1975). Mammalian DNA polymerases. Prog Nucleic Acids Res 15:109.

Buckler-White AJ, Humphrey GW, Pigiet V (1980). Association of polyoma T antigen and DNA with the nuclear matrix from lytically infected 3T3 cells. Cell 22:37.

Cook PR, Brazell IA (1975). Supercoils in human DNA. J Cell Sci 19:261.

Dijkwel PA, Mullenkers LHF, Wanka F (1979). Analysis of the attachment of replicating DNA to a nuclear matrix in mammalian interphase nuclei. Nucleic Acids Res 6:219.

Fakan S, Bernhard W (1971). Localization of rapidly and slowly labeled nuclear RNA as visualized by high resolution autoradiography. Exp Cell Res 67:129.

Fakan S, Hancock R (1974). Localization of newly synthesized DNA in a mammalian cell as visualized by high resolution autoradiography. Exp Cell Res 83:95.

Fakan S, Puvion E, Spohr G (1976). Localization and characterization of newly synthesized nuclear RNA in isolated rat hepatocytes. Exp Cell Res 99:155.

Georgiev GP, Nedospasov SA, Bakayev VV (1978). Supranucleosomal levels of chromatin organization. Cell Nucleus 6:3.

Hemminki K, Vainio H (1979). Preferential binding of benzo-(α)pyrene into nuclear matrix fraction. Cancer Lett 6:167.

Hodge LD, Mancini P, Davis FM, Heywood P. (1977). Nuclear matrix of HeLa S_3 cells. Polypeptide composition during adenovirus infection and in phases of the cell cycle. J Cell Biol 72:194.

Igo-Kemenes T, Zachau HG (1978). Domains in chromatin structure. Cold Spring Harbor Symp Quant Biol 43:109.

Jeppesen PGN, Bankier AT (1979). A partial characterization of DNA fragments protected from nuclease degradation in histone depleted metaphase chromosomes of the Chinese hamster. Nucleic Acids Res 7:49.

Lenk R, Ransom L, Kaufman Y, Penman S (1977). A cytoskeletal structure with associated polyribosomes obtained from HeLa cells. Cell 10:67.

Long BH, Huang C, Pogo AO (1979). Isolation and characterization of the nuclear matrix in Friend erythroleukemia cells: chromatin and hn-RNA interactions with the nuclear matrix. Cell 18:1079.
Osborn M, Weber K (1977). The detergent resistant cytoskeleton of tissue culture cells includes the nucleus and microfilament bundles. Exp Cell Res 106:339.
Pardoll DM, Vogelstein B, Coffey DS (1980). A fixed site of DNA replication in eucaryotic cells. Cell 19:527.
Paulson JR, Laemmli UK (1977). The structure of histone-depleted metaphase chromsomes. Cell 12:817.
Razin SV, Mantieva VL, Georgiev GP (1978). DNA adjacent to attachment points to deoxyribonucleoprotein fibrils to chromosomal axial structure is enriched in reiterated base sequences. Nucleic Acids Res 5:4737.
Razin SV, Mantieva VL, Georgiev GP (1979). The similarity of DNA sequences remaining bound to scaffold upon nuclease treatment of interphase nuclei and metaphase chromosomes. Nucleic Acids Res 7:1713.
Smith HC, Berezney R (1980). DNA polymerase α is tightly bound to the nuclear matrix of actively replicating liver. Biochem Biophys Res Commun 97:1541.
Smith HC, Berezney R (1981). In vitro DNA synthesis in isolated nuclear matrix. J Cell Biol in press.
Swift H (1963). Cytochemical studies of nuclear fine structure. Exp Cell Res Suppl 9:54.
van Eekelen CAG, van Venrooij WJ (1981). Hn-RNA and its attachment to a nuclear protein matrix. J Cell Biol 88:554.
Wolosewick JJ, Porter KR (1976). Stero high voltage electron microscopy of whole cells of the human diploid cell line WI-38. Am J Anat 147:303.
Wolosewick JJ, Porter KR (1979). Microtrabecular lattice of the cytoplasmic ground substance. Artifact or reality? J Cell Biol 182:114.
Wunderlich F, Berezney R, Kleinig H (1976). The nuclear envelope: an interdisciplinary analysis of its morphology, composition and functions. In Chapman D, Wallach DFH (eds): "Biological Membranes," Vol 3, London: Academic Press, p 241.

THE ATTACHMENT OF REPLICATING DNA TO THE NUCLEAR MATRIX

Friedrich Wanka, Anna C.M. Pieck, Ad G.M. Bekers
and Leon H.F. Mullenders

Laboratory of Chemical Cytology
University of Nijmegen
Nijmegen, The Netherlands

Recent studies have revealed that the nuclear matrix plays a role in the spatial organization of DNA replication (Wanka et al. 1977; Dijkwel et al. 1979; Pardoll et al. 1980; and others). We became interested in the problem by the consideration that a structural component must be involved to unwind the DNA double helix in such a way that the daughter molecules can separate unimpededly in the subsequent mitosis. How this can be achieved is shown diagrammatically in Figure 1 (see also Dingman 1974). Stage A represents a loop of a DNA molecule attached to a scaffold by two successive origins of replications. It is supposed that the origins and attachment sites are duplicated when DNA synthesis starts and that soon afterwards the replication forks become bound to additional attachment sites (stage B). The attachment of the replication fork must be of such a nature that the DNA molecule is completely unwound when it moves through the binding site on the scaffold (stage C and D). For the complete unwinding, it is important that the binding sites lie close together and remain in a fixed orientation with regard to each other until the replication is completed.

THE 2 M NaCl-RESISTANT DNA-PROTEIN COMPLEX IN MAMMALIAN CELLS

In the search for a nuclear scaffold we isolated nuclei from an established line of calf liver cells cultured in vitro (Pieck 1971). Triton X-100 (0.1%) was included in the isolation buffer in order to remove the nuclear membranes (Wanka et al. 1977). The chromatin of the isolated nuclei

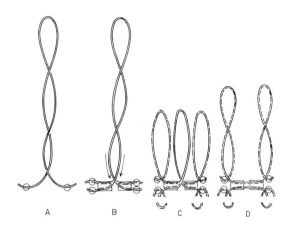

Fig. 1. Hypothetical attachment of the DNA to a nuclear scaffold during replication. Circles and boxes represent attachment sites for origins of replication and replication forks, respectively. The interrupted line represents the newly synthesized DNA strand. A: Attachment of two successive origins of a non-replicating DNA molecule. B: Initiation of the replication by duplication of the origins and origin attachment sites; attachment of the replication forks. C: Translocation of the DNA molecule through the replication binding site. The direction is indicated by the arrows in B. D: Termination of the replication resulting in two fully untangled DNA daughter molecules.

was dissociated by addition of NaCl at a final concentration of 2 M. Such a nuclear lysate was prepared from cells labeled for 40 hr with [^{3}H]leucine and [^{14}C]thymidine, and analyzed by centrifugation through a sucrose gradient. Experimental details are described elsewhere (Wanka et al. 1977). We found a rapidly sedimenting component, consisting of 15 to 20% of the nuclear protein and between 50 and 100% of the DNA. The 2 M NaCl-resistant nuclear structure can be dissolved by addition of 5 M urea to the lysate. At suitable centrifugation conditions, DNA and protein sediment as two separate peaks (Fig. 2).

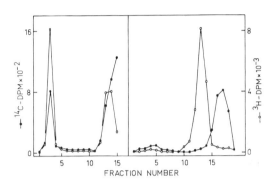

Fig. 2. Demonstration of the 2 M NaCl-resistant DNA-protein complex and its dissociation by urea. A nuclear lysate in 2 M NaCl was prepared and solid urea was added at a final concentration of 5 M to one part. The samples were then centrifuged at 25,000 rpm through 30 to 40% sucrose gradients. Left panel: 2 M NaCl control (centrifuged for 1 hr). Right panel: 5 M urea added (centrifuged for 16 hr).
(o) [^{14}C]-labeled DNA. (●) [^{3}H]-labeled protein. Sedimentation is from right to left.

The protein composition of the salt-resistant nuclear structure has been compared with nuclear and cytoplasmic proteins by sodium dodecyl sulfate (SDS)-polyacrylamide gel electrophoresis. It shows a number of specific polypeptide bands (Fig. 3) known already from nuclear matrix proteins (Berezney, Coffey 1974, 1977). The three major bands with apparent molecular weights of 71, 66 and 60 kilodaltons have been further identified as lamins (Gerace, Blobel 1980; Shelton et al. 1980) by isoelectric focusing.

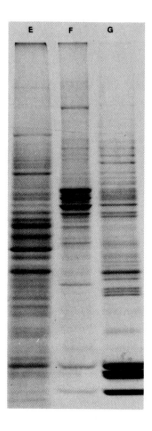

Fig. 3. PAGE of the polypeptides obtained from the 2 M NaCl-resistant DNA-protein complex. The preparation and electrophoresis was as described by Mitchelson et al. (1979). E, cytoplasm; F, DNA-protein complex; G, nuclei.

DNA ATTACHMENT TO THE MATRIX IN MAMMALIAN CELLS

Investigation by electron microscopy reveals that the rapidly sedimenting material consists of residual nuclear structures which contain major elements of the nuclear matrix, namely, the lamina with residual pore complexes and some morphologically less well defined internal material. Large amounts of DNA threads are distributed throughout the entire matrix and a zone around it (Fig. 4) which is comparable to the halo observed by light microscopy with fluorescent dyes (Vogelstein et al. 1981). Some insight into the DNA binding was obtained by partial digestion with DNase.

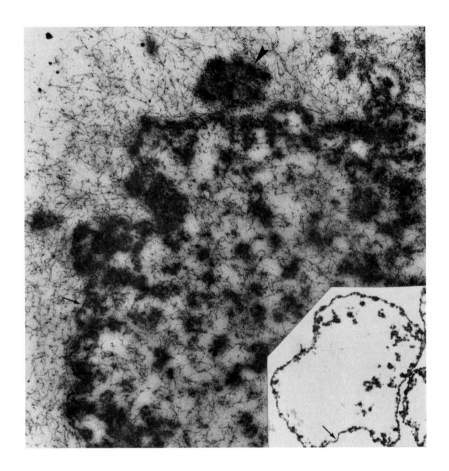

Fig. 4. Thin-section electron micrograph of the 2 M NaCl-resistant DNA-protein complex. Nuclear lamina (arrow). Residual pore complexes (arrowhead), to be seen in tangential sections. Preparation as described by Bekers et al. (1981). Magnification 33,000X. <u>Inset:</u> residual NaCl-resistant nuclear structure after exhaustive digestion by DNase. Magnification 6400X.

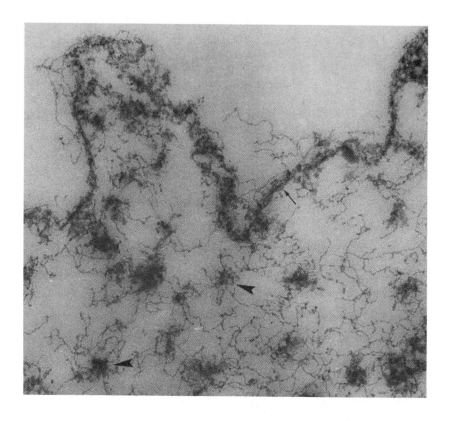

Fig. 5. Thin section of the 2 M NaCl-resistant structure after digestion with 0.2 U/ml DNase for 30 min. DNA threads associated with the lamina (arrow) and internal matrix (arrowheads). Magnification 84,000X.

The short remaining DNA threads appear associated with both the lamina and the internal matrix elements (Fig. 5). We also spread matrix fragments, produced by shearing, with the Kleinschmidt technique. In such fragments, DNA threads could be seen to emerge from the residual annular structures of the nuclear pores. No DNA threads were visible when the matrix material was extensively digested with DNase. In addition, the internal matrix collapsed when the DNA was removed (Fig. 4, inset). This contrasts with the finding of a stable internal matrix by other authors in mammalian cells (Berezney, Coffey 1974, 1977; Comings, Okada 1976) and other organisms (Herlan, Wunderlich 1976; Mitchelson et al. 1979, Bekers et al. 1981). We have some evidence that the reduced stability of the internal matrix and the loss of the residual nucleolus are due to the omission of Mg^{++} from the isolation medium.

ASSOCIATION OF THE NEWLY REPLICATED DNA WITH THE MATRIX IN PHYSARUM

It has been reported repeatedly that, in mammalian cells, replicating DNA is attached to the nuclear matrix by a region close to the replication fork (Wanka et al. 1977; Dijkwel et al. 1979; Pardoll et al. 1980; Vogelstein et al. 1981; McCready et al. 1980). This binding is of a temporary nature. Little is known of permanent binding sites. Cook and Brazell (1981) have observed a binding site close to the α-globin gene locus. Polyoma DNA also becomes bound to the nuclear matrix. The binding site appears to be close to the origin of replication of the viral DNA (Buckler-White et al. 1980).

To demonstrate the binding of the origins of replicons requires the specific labeling of the origin regions. A partial labeling of the origins can be achieved in plasmodia of Physarum polycephalum. The nuclei of Physarum plasmodia divide synchronously every 7-8 hr, followed immediately by the S phase (Nygaard et al. 1960). To determine the duration of the S phase, a plasmodium was prelabeled for 17 hr with $[^{14}C]$thymidine and, starting at mitosis 3, small fragments were cut off and pulse-labeled for 10 min with $[^3H]$thymidine. As shown in Figure 6, the S phase starts 5-10 min after metaphase by an abrupt increase of the thymidine incorporation and ends about 2 hr later. This synchrony makes it possible to study the attachment of the DNA synthesized at different times during the S phase. In particular, it seems possible to preferentially label origins of replication at the very beginning of the S phase, and DNA regions remote from the origins at the end of the S phase.

We studied the binding of the DNA regions to the matrix by controlled digestion with DNase of nuclear lysates in 2 M NaCl obtained from pulse-labeled plasmodia. The digested

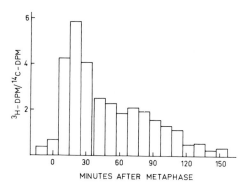

Fig. 6. Time course of thymidine incorporation into sections of a Physarum plasmodium during the S phase. Bars represent the amount of ^{3}H incorporated per unit of [^{14}C]-labeled bulk DNA per 10 min.

material was then analyzed by centrifugation in sucrose gradients. A typical experiment with a plasmodium, pulse-labeled from 15 to 20 min after metaphase, is shown in Figure 7. The bulk DNA, represented by ^{14}C counts, is released from the matrix with increasing DNase concentrations. Newly synthesized [^{3}H]-labeled DNA, on the other hand, is much less prone to the DNase action. Thus, the ratio of newly synthesized to bulk DNA increases with decreasing amounts of bulk DNA remaining associated with the matrix (Fig. 7). This result has been confirmed in pulse experiments carried out at different times of the S phase (Fig. 8, upper line). In all experiments the ratio of replicating DNA to bulk DNA, as indicated by the ^{3}H/^{14}C ratio, increased with decreasing amounts of DNA remaining at the matrix. Similar results have already been reported (Wanka, Mitchelson 1979; Hunt, Vogelstein 1981). Obviously, a DNA region located close to the replication fork is attached to the matrix.

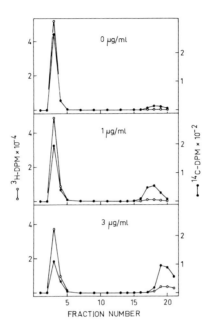

Fig. 7. Association of newly replicated DNA with the nuclear matrix. A Physarum plasmodium, prelabeled with 0.2 µCi/ml [^{14}C]dThd, was pulse-labeled for 5 min with 50 µCi/ml [^{3}H]dThd, starting 15 min after metaphase. A nuclear lysate was prepared in 2 M NaCl and samples were digested for 15 min with DNase I at the concentrations indicated. The lysate was then centrifuged for 1 hr at 20,000 rpm through 15 to 40% sucrose gradients. The fractions were analyzed as described previously (Wanka et al. 1977).

ASSOCIATION OF THE ORIGINS OF REPLICONS WITH THE MATRIX IN PHYSARUM

If the replication forks were the only attachment sites, one might expect no DNA to be attached to the matrix in the G_2 phase. This is not so. In lysates prepared in the G_2 phase, we always found a proportion of the DNA associated with the matrix, similar to that observed in the S phase. To find out whether this is a random association, we chased pulse-labeled plasmodia into late G_2 phase and analyzed the distribution of the pulse label. Pulse label given early in the S phase and chased into late G_2 phase was not specifically incorporated in either the attached regions or the loop regions of the DNA molecules. The $^3H/^{14}C$ ratio of the DNA remaining after various DNase digestions did not change significantly (Fig. 8). However, pulse label given in the late S phase was more efficiently removed from the matrix by DNase than bulk DNA as can be inferred from the decreased $^3H/^{14}C$ ratio in the rapidly sedimenting material (lower line in Fig. 8). Apparently, in late S phase thymidine is incorporated in a position remote from the G_2 phase binding sites. This is compatible with the hypothesis that the G_2 binding sites are at the origins of replication.

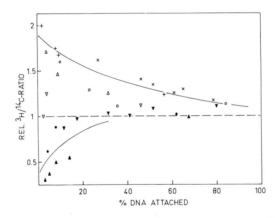

Fig. 8. Association of pulse-labeled DNA with the matrix during the nuclear cycle. Prelabeled plasmodia of <u>Physarum</u> were pulse-labeled at various times during the S phase and analyzed as described in Figure 7, either immediately or after a chase into late G_2 phase (6 hr after metaphase). Relative ratios (% ^3H/% ^{14}C) of the matrix-associated DNA are recorded as functions of the DNA (% ^{14}C) remaining associated with the matrix at various DNase treatments. Labeling times of the pulse experiments (min after metaphase): + 35-40, o 55-60, ∆ 80-85, x 100-110; chase experiments: ∇ 35-40, ● 55-60, ▼ 75-85, ▲ 100-110.

The binding of the origins of replicons was confirmed by a final experiment in which plasmodia were labeled from prophase to 10 min after metaphase and then chased until 6 hr after metaphase. This results in a specific labeling of the origins including short adjacent regions. Immediately after the pulse, the label was again found to be preferentially associated with the matrix material (Fig. 9). In this particular experiment, however, the pulse label remained preferentially associated with the matrix until the late G_2 phase. Our tentative interpretation of these results is that the origins of early replicons, and possibly of all replicons, are permanently bound to the nuclear matrix.

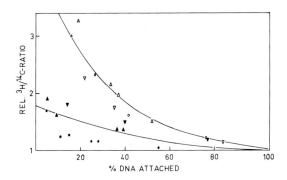

Fig. 9. Association of the origins of replicons with the matrix in S phase and G_2 phase. Prelabeled plasmodia were pulse-labeled from prophase to 10 min after metaphase (o, △, ▽). One half of each plasmodium was chased until 6 hr after metaphase (●, ▲, ▼). (Three independent experiments; further details as in Figure 8.)

SUMMARY AND CONCLUSIONS

The results suggest that nuclear DNA is attached to three elements of the nuclear matrix, i.e. the lamina, the pore complexes and the internal matrix. Permanently bound DNA regions appear to be the origins of replicons. Additional binding occurs, during DNA synthesis, at the replication forks. Whether the binding regions on the DNA molecule correspond to specific attachment sites on the matrix is not known.

Binding of the origins of replication and of the replication forks has been postulated in a model to explain the unwinding of the DNA double helix during the replication (Dingman 1974). This model requires that the attachment sites for both regions lie close together, preferentially at the nuclear lamina or pore complexes. However, available data on the spatial intra-nuclear location of the initiation and replication processes are not conclusive on this point (O'Brien et al. 1972; Hanania, Harel 1973; Fakan, Hancock 1974).

ACKNOWLEDGEMENT

We would like to thank A.A.M. Rijken and J.M.A. Aelen for excellent technical assistance with the experiments.

Bekers AGM, Gijzen HJ, Taalman RDFM, Wanka F (1981). Ultrastructure of the nuclear matrix from Physarum polycephalum during the mitotic cycle. J Ultrastruct Res, in press.
Berezney R, Coffey DS (1974). Identification of a nuclear protein matrix. Biochim Biophys Res Commun 60:1410.
Berezney R, Coffey DS (1977). Nuclear matrix: isolation and characterization of a framework structure from rat liver nuclei. J Cell Biol 73:616.
Buckler-White AJ, Humphrey GW, Pigiet V (1980). Association of polyoma T antigen and DNA with the nuclear matrix from lytically infected 3T6 cells. Cell 22:37.
Comings DE, Okada TA (1976). Nuclear proteins. III. The fibrillar nature of the nuclear matrix. Exp Cell Res 103:341.
Cook PR, Brazell IA (1980). Mapping sequences in loops of nuclear DNA by progressive detachment from the nuclear cage. Nucl Acids Res 8:2895.
Dijkwel PA, Mullenders LHF, Wanka F (1979). Analysis of the attachment of replicating DNA to a nuclear matrix in mammalian interphase nuclei. Nucl Acids Res 6:219.
Dingman CW (1974). Bidirectional chromosome replication: some topological considerations. J Theor Biol 43:187.
Fakan S, Hancock R (1974). Localization of newly synthesized DNA in mammalian cells as visualized by high resolu-

tion autoradiography. Exp Cell Res 83:95.
Gerace L, Blobel G (1980). The nuclear envelope lamina is reversibly depolymerized during mitosis. Cell 19:277.
Hanania N, Harel J (1973). Replication de l'ADN et membrane nucleaire dans les cellules animales. Biochemie 55:357.
Herlan G, Wunderlich F (1976). Isolation of a nuclear protein matrix from Tetrahymena macronuclei. Cytobiologie 13:291.
Hunt BF, Vogelstein B (1981). Association of newly replicated DNA with the nuclear matrix of Physarum polycephalum. Nucl Acids Res 9:349.
McCready SJ, Godwin J, Mason DW, Brazell IA, Cook PR (1980). DNA is replicated at the nuclear cage. J Cell Sci 46:365.
Mitchelson KR, Bekers AGM, Wanka F (1979). Isolation of a residual protein structure from nuclei of the myxomycete Physarum polycephalum. J Cell Sci 39:247.
Nygaard OF, Guttes S, Busch H (1960). Nuclei acid metabolism in a slime mold with synchronous mitosis. Biochim Biophys Acta 38:298.
O'Brien RL, Sanyal AB, Stanton RH (1973). DNA replication sites in HeLa cells. Exp Cell Res 80:340.
Pardoll DM, Vogelstein B, Coffey DS (1980). A fixed site of DNA replication in eukaryotic cells. Cell 19:527.
Pieck ACM (1971). Extraction, separation and identification of the free nucleotides in liver cells cultivated in vitro. Proc Kon Ned Acad Wet Ser C 74:303.
Shelton KR, Higgins LL, Cochran DL, Ruffolo Jr JJ, Egle PM (1980). Nuclear lamins of erythrocytes and liver. J Biol Chem 255:10978.
Vogelstein B, Pardoll DM, Coffey DS (1980). Supercoiled loops and eucaryotic DNA replication. Cell 22:79.
Wanka F, Mitchelson KR (1979). Attachment of the nuclear DNA to a high salt resistant nuclear structure in Physarum polycephalum. In Sachsenmayer W (ed): "Current Research on Physarum," Vol 120, Innsbruck: University of Innsbruck Press, p 59.
Wanka F, Mullenders LHF, Bekers AGM, Pennings LJ, Aelen JMA, Eygensteyn J (1977). Association of nuclear DNA with a rapidly sedimenting structure. Biochem Biophys Res Commun 74:739.

tion autoradiography. Exp Cell Res 83:95.
Gerace L, Blobel G (1980). The nuclear envelope lamina is reversibly depolymerized during mitosis. Cell 19:277.
Hanania N, Harel J (1973). Replication de l'ADN et membrane nucleaire dans les cellules animales. Biochemie 55:357.
Herlan G, Wunderlich F (1976). Isolation of a nuclear protein matrix from Tetrahymena macronuclei. Cytobiologie 13:291.
Hunt BF, Vogelstein B (1981). Association of newly replicated DNA with the nuclear matrix of Physarum polycephalum. Nucl Acids Res 9:349.
McCready SJ, Godwin J, Mason DW, Brazell IA, Cook PR (1980). DNA is replicated at the nuclear cage. J Cell Sci 46:365.
Mitchelson KR, Bekers AGM, Wanka F (1979). Isolation of a residual protein structure from nuclei of the myxomycete Physarum polycephalum. J Cell Sci 39:247.
Nygaard OF, Guttes S, Busch H (1960). Nuclei acid metabolism in a slime mold with synchronous mitosis. Biochim Biophys Acta 38:298.
O'Brien RL, Sanyal AB, Stanton RH (1973). DNA replication sites in HeLa cells. Exp Cell Res 80:340.
Pardoll DM, Vogelstein B, Coffey DS (1980). A fixed site of DNA replication in eukaryotic cells. Cell 19:527.
Pieck ACM (1971). Extraction, separation and identification of the free nucleotides in liver cells cultivated in vitro. Proc Kon Ned Acad Wet Ser C 74:303.
Shelton KR, Higgins LL, Cochran DL, Ruffolo Jr JJ, Egle PM (1980). Nuclear lamins of erythrocytes and liver. J Biol Chem 255:10978.
Vogelstein B, Pardoll DM, Coffey DS (1980). Supercoiled loops and eucaryotic DNA replication. Cell 22:79.
Wanka F, Mitchelson KR (1979). Attachment of the nuclear DNA to a high salt resistant nuclear structure in Physarum polycephalum. In Sachsenmayer W (ed): "Current Research on Physarum," Vol 120, Innsbruck: University of Innsbruck Press, p 59.
Wanka F, Mullenders LHF, Bekers AGM, Pennings LJ, Aelen JMA, Eygensteyn J (1977). Association of nuclear DNA with a rapidly sedimenting structure. Biochem Biophys Res Commun 74:739.

THE ROLE OF THE NUCLEAR MATRIX IN POLYOMA DNA REPLICATION

Vincent Pigiet and Glen W. Humphrey

Department of Biology
The Johns Hopkins University
Baltimore, Maryland 21218

Polyoma virus contains a DNA genome that is small (3.6 kilodaltons (K), 5200 bp), affording only a limited coding capacity for only a few proteins (Tooze 1980). The early transcription class includes the three tumor (T) antigens -- the big T antigen (100K), middle T antigen (~60K), and little T antigen (22K) -- which share a common N terminus, but differ in their internal sequences due to the different modes of RNA splicing (Tooze 1980). The late transcription class includes the three coat proteins, each of which is coded for from overlapping regions of the genome. Both genetic and biochemical data have implicated the 100K T antigen in regulating transcription levels of both early and late messages, as well as regulating the initiation of each round of DNA replication. Each of these functions is thought to be mediated by the number and arrangement of T antigen promoters binding to the region of the genome defined as the origin of replication (Tjian 1978; Meyers et al. 1981; McKay 1981).

The limited genetic content of the polyoma genome dictates that the host cell must supply the vast array of enzymes structural proteins, and nucleotide precursors required for replication. Given this intimate dependence, the mode of replication of the viral genome is very likely to reflect closely the replication mode of the host genome. Technically, analysis of the replication mechanism of the virus is greatly aided by the small physical size of the DNA and the certain definition of the T antigen as the regulatory protein controlling initiation of replication and gene activation.

VIRAL DNA AND THE NUCLEAR MATRIX

In earlier studies on the role of the nuclear matrix on polyoma replication, we described a procedure for isolation of the matrix from virus-infected cells that would effectively remove the bulk of the cellular DNA without damaging the circular viral genomes remaining on the matrix (Buckler-White et al. 1980). Extending these studies we have used a variety of hybridization approaches to show that residual viral DNA represented approximately 6% of the total viral DNA at a time when the rate of viral DNA synthesis was maximal (~26 hr post-infection). At earlier times after infection (<20 hr) when viral replication begins, the percentage of viral DNA on the matrix is even greater, amounting to 25% at 16 hr post-infection. Pulse labeling experiments in vivo which allow discrimination of the actively replicating molecules (i.e., 5 min pulse label) showed that throughout the period of active replication (20-28 hr) the majority (~80%) of the viral replicating molecules is matrix bound. Likewise, labeling for 30 min which accounts for several rounds of replication, the matrix-bound fraction is greater than 70% (Buckler-White and Pigiet, unpublished data).

By a variety of criteria including electrophoretic fractionation of replicative intermediates and restriction nuclease analysis, no preference was observed for replicating molecules at any particular stage of replication. These findings imply that the entire process of replication from initiation through termination and progeny segregation may occur through matrix-associated events.

NATURE OF THE T ANTIGEN ASSOCIATED WITH THE NUCLEAR MATRIX

Previous studies showed that the 100K T antigen in lytically infected 3T6 cells exists in either a free (or soluble) form extractable with non-ionic detergent or in a form tenaciously bound to the nuclear matrix (Buckler-White et al. 1980). Since the soluble species was both abundant and immunoreactive, we could quantify the accumulation of T antigen (in µg) after electrophoresis of immune precipitates by comparison with a protein standard, bovine serum albumin. Direct quantitation of the matrix-bound T antigen was not possible in these studies because of detectability limits and because the sodium dodecyl sulfate (SDS)-solubilized protein was no longer antigenic using anti-tumor sera. In-

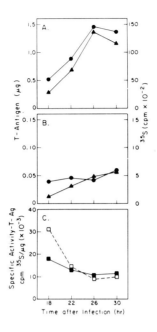

Fig. 1. Accumulation and [^{35}S]-labeling of viral T antigen in cell extracts and on nuclear matrix as a function of time after infection. Three plates of polyoma-infected cells were labeled for 2 hr with [^{35}S]methionine and [^{32}P]orthophosphate prior to harvest. Cell extracts (one plate) and nuclear matrix (two plates) were prepared at 18, 22, 26, and 30 hr after infection. Immunoprecipitates of cell extracts and solubilized nuclear matrix samples were electrophoresed on 8% SDS-acrylamide gels, stained for protein and Coomassie dye, dried and autoradiographed. ^{35}S or ^{32}P cpm incorporated was determined by excision and liquid scintillation counting of the 100K band accumulation of T antigen in cell extracts by quantitative densitometry of the Coomassie stained gel and accumulation of T antigen or nuclear matrix by crossed immunoelectrophoresis (Converse, Papermaster 1975; Humphrey et al., unpublished results). Cell extracts (A) and matrix samples (B) were analyzed for 100K T antigen in μg (-▲-) and cpm of [^{35}S]methionine (-●-). Specific activity in cpm/μg (C) was calculated from the data of (A) and (B) for T antigen cell extracts (-■-) and nuclear matrix samples (-□-).

direct methods, however, showed that the 100K T antigen was present on matrix and that the pattern of kinetic labeling with [^{35}S]methionine was different from that for the soluble species. The recent development of an antibody reactive to the denatured protein, combined with the technique of crossed-rocket immunoelectrophoresis has now allowed us to quantify both classes of T antigen free from ambiguities inherent in assumptions of equivalent isotope labeling of compartmentalized proteins (Humphrey et al., unpublished results).

As shown in Figure 1, both the soluble and the matrix-bound forms of the 100K T antigen accumulated as infection progressed, approximately following the rate of polyoma DNA synthesis (Buckler-White et al. 1980). The relative amount of T antigen bound to matrix was approximately constant at 6% through this infection time course, except at early times (< 20 hr) when the fraction bound to matrix was substantially higher. The relative specific radioactivity (i.e., cpm ^{35}S/µg T antigen) showed that for the long labeling period used (2 hr), the two classes of T antigen were comparable in their kinetic labeling patterns, except at early times when the matrix-bound class was more highly labeled.

T ANTIGEN BOUND TO THE NUCLEAR MATRIX IS HIGHLY PHOSPHORYLATED

Use of this quantitative methodology has now allowed us to determine the molar ratio of phosphate bound to the two classes of T antigen. This analysis was carried out on infected cell cultures labeled in vivo with [^{32}P]orthophosphate and [^{35}S]methionine for 2 hr just prior to harvest at 26 hr after infection. Parallel samples containing the soluble and the matrix-bound classes of T antigen from each set of samples were immunoprecipitated and electrophoresed

Fig. 2 (following pages). Synthesis, accumulation and phosphorylation of the polyoma T antigen (100K) in cell extracts and on the nuclear matrix. Immunoprecipitate anti-T or preimmune, prepared as described in legend to Figure 1, electrophoresed on 8% SDS polyacrylamide gels (B., D. 1 day, D., E. 6 days). (A) Cell extract, Coomassie-stained; (B) as in (A) autoradiography: ^{35}S + ^{32}P; (C) Nuclear matrix; autoradiography; ^{35}S + ^{32}P; (D) Cell extract, autoradiography: ^{32}P only; (E) Nuclear matrix autoradiography: ^{32}P alone.

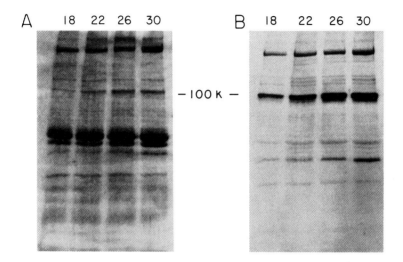
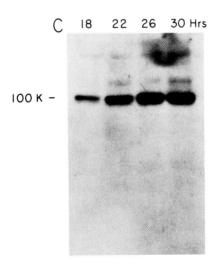

Fig. 2

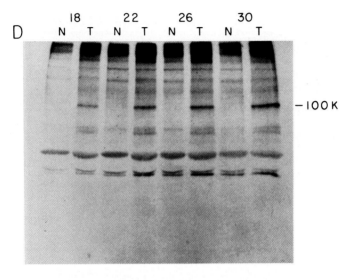

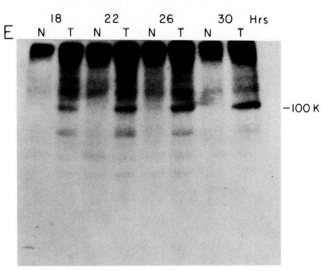

Fig. 2

on SDS-polyacrylamide gel electrophoresis (PAGE). Sufficient T antigen in the soluble class was produced throughout the entire time course to allow easy visualization by direct staining with Coomassie blue dye (Fig. 2A). Autoradiography of this gel detecting ^{32}P and ^{35}S (Fig. 2B) or ^{32}P alone (Fig. 2C) shows that the soluble form of T antigen was phosphorylated throughout the infection time course. Comparable autoradiographs for the nuclear matrix samples (Figs. 2D, E) also show phosphorylation of T antigen. Since the matrix-bound T antigen was undetectable by direct protein staining we quantified this amount by the crossed-rocket technique (Converse, Papermaster 1975; Humphrey, Pigiet unpublished results). Using the determined specific radioactivity of [^{32}P]phosphate pool we could determine the molar ratio of phosphate per mole of 100K T antigen (Fig. 3). Two observations are notable: First, the molar ratios are high, approximately constant at 5 moles of phosphate per mole of protein for the soluble species and ranging from 10 to 25 for the matrix-bound species. Second, the relative phosphorylation of the matrix-bound species was higher at all times after infection with the greatest difference at early times

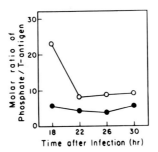

Fig. 3. Molar ratio of phosphate to viral T antigen as a function of time after infection. The accumulation and ^{32}P labeling of T antigen were calculated from data presented in Figures 2 and 3 as described in legend to Figure 2. Moles of phosphate were calculated from ^{32}P cpm incorporated in vivo into T antigen isolated either in cell extracts (-●-) or nuclear matrix samples (-o-) using the specific activity of total phosphate in the cell extracts.

after infection. These conclusions are consistent with earlier studies reported in the relative labeling in vivo with [^{32}P]orthophosphate and [^{35}S]methionine (^{32}P/^{35}S) for the two classes of T antigens (Buckler-White et al. 1980).

The unexpectedly high molar ratios of phosphate bound to T antigen suggested a polymeric form of phosphate, probably contained in a diester linkage. This idea was confirmed by the sensitivity of the ^{32}P in 100K T antigen to snake venom diesterase. Experiments carried out with the soluble form of T antigen obtained after immunoprecipitation and gel electrophoresis showed that approximately one-third of the ^{32}P was released by diesterase. This released fraction behaved like the expected digestion products of poly ADP ribose by the criteria of thin-layer chromatography (Pekala et al. 1981).

SUMMARY AND CONCLUSIONS

The nuclear matrix has been implicated in the replication of mammalian DNA (Berezney, Coffey 1975; Pardoll et al. 1980), and for DNA of the virus polyoma, which produces a lytic response in susceptible mouse cells (Buckler-White et al. 1980). Evidence for all stages of replication taking place on the matrix was obtained by the isolation of the entire spectrum of viral replication intermediates in the matrix. Early in the lytic cycle when the majority of genomes present are involved in active replication, a higher fraction is observed for matrix-associated DNA.

The parallel association with the matrix of the initiation control protein, the T antigen, suggested a functional role of matrix in initiation, and possibly control of transcription. The high level of phosphorylation of T antigen, as well as the differences in the phosphorylation of the matrix-bound species, may provide a basis for proposing a post-translational modification of T antigen with poly ADP ribose in facilitating binding to matrix and in directing the pleiotropic roles for T antigen in replication and transcriptional control.

Berezney R, Coffey DS (1975). Nuclear protein matrix: association with newly synthesized DNA. Science 189:291.

Buckler-White AJ, Humphrey GH, Pigiet V (1980). Association of polyoma T antigen and DNA with the nuclear matrix from lytically infected 3T6 cells. Cell 22:37.

Converse CA, Papermaster DS (1975). Membrane protein analysis by two-dimensional immunoelectrophoresis. Science 189:469.

McKay RDG (1981). Binding of a Simian Virus 40 T antigen-related protein to DNA. J Mol Biol 145:471.

Meyers RM, Rio DC, Robbins AK, Tjian R (1981). SV40 gene expression is modulated by the cooperative binding of T antigen to DNA. Cell 25:373.

Pardoll PM, Vogelstein B, Colby DS (1980). A fixed site of DNA replication in eukaryotic cells. Cell 19:527.

Pekala PH, Lane MD, Watkins PA and Moss J (1981). On the mechanism of preadipocyte differentiation: masking of poly (ADP-ribose) synthetase activity during differentiation of 3T3-L1 preadipocyte. J Biol Chem 256:4871.

Tjian R (1978). The binding site on SV40 DNA for a T antigen-related protein. Cell 13:165.

Tooze J (1980). "The Molecular Biology of Tumor Viruses, DNA Tumor Viruses", 2nd edition, New York: Cold Spring Harbor Press.

CROSSLINKING EXPERIMENTS IN NUCLEAR MATRIX: NONHISTONE PROTEINS TO HISTONES AND SnRNA TO HnRNA

A. Oscar Pogo, M.D., D.M. Sci, Luis Cornudella, Ph.D.,* Alice E. Grebanier, Ph.D., Roman Procyck, Ph.D., and Valerie Zbrzezna, B.S.

Laboratory of Cell Biology
The Lindsley F. Kimball Research Institute of the
 New York Blood Center
310 East 67th Street
New York, NY 10021

During the course of experiments to elucidate the nature of the nuclear matrix, it was discovered that the morphology of the nucleus is preserved after almost all of the DNA is removed by DNase I digestion (Long et al. 1979). The intranucleosomal histones (H2a, H2b, H3 and H4) remained within the nuclear matrix after digestion of the DNA, and as monitored by the use of crosslinking reagents, the interactions between histones appeared to be undisturbed (Long et al. 1979; Grebanier, Pogo 1979). The histones are known to be organized into octameric units, the nucleosomal cores, around which the DNA is wrapped (Kornberg 1974; Van Holde et al. 1974). However, they normally do not maintain such large structures at physiological ionic strength without DNA. Therefore, the interactions between the histones and the other nuclear proteins must be responsible for the preservation of nuclear morphology following DNase I treatment. In addition, nuclear particles that contain hnRNA (hnRNP), nuclear particles that contain snRNA (snRNP), ribosomal subunits and ribosomal precursors are maintained within the nucleus in the absence of DNA. Treatments with either low- or high-salt buffers produce no release of

---

*Visiting Scientist from the Instituto de Quimica Bio-Organica del C.S.I.C., Universidad Politecnica, Barcelona, Spain.

these components (Long et al. 1979). Therefore, the interactions between these structures and a highly ordered, supramolecular structure insoluble in either high or low salt must be responsible for maintaining heterogeneous RNA (hnRNA), pre-rRNA, small nuclear RNA (snRNA) and rRNA within the nucleus. These observations suggested that removal of DNA is not accompanied by any major rearrangement of nuclear structures.

As a first approach to investigate the nuclear architecture, crosslinking experiments were performed in whole nuclei and in DNA-depleted nuclei. These experiments revealed a set of nonhistone proteins that seemed able to form disulfide bonds with H3 when the nuclei were oxidized with H_2O_2 (Grebanier, Pogo 1979, 1981). These nonhistone proteins seem to be in close association with H3, and they are of particular interest in light of their possible role in forming a link between the histones and the nuclear matrix. In addition, the interaction of nonhistone proteins with H3 is particularly important since H3, like H4, plays a pivotal role in nucleosome assembly (Jackson, Chalkley 1981). Further, covalent linkages had been formed between a set of snRNA and hnRNA when DNA-depleted nuclei were irradiated with UV light (352 nm) in the presence of 4'aminomethyl-4,5'8-trimethyl psoralen (AMT). This observation suggests that a set of snRNA interacts with hnRNA in a sequence-specific manner in the nucleus, and that these snRNA may be involved in processing and perhaps transport of mRNA.

NONHISTONE PROTEINS CROSSLINKED BY DISULFIDE BONDS TO H3

This study was undertaken using undifferentiated and differentiated Friend erythroleukemia cells (Long et al. 1979). To this end, disulfide bonds were formed between proteins in nuclei by oxidation with H_2O_2 and the histones extracted either as nucleosomes from micrococcal nuclease-treated samples, or as DNA-free proteins from DNase I-treated samples. Particular crosslinked, nonhistone proteins were extracted together with histones by these two procedures (Grebanier, Pogo 1981).

The extracts were examined for crosslinked proteins by two-dimensional electrophoresis (Wang, Richards 1974). The first electrophoresis step was done without reducing reagents but in the second, β-mercaptoethanol was added. Polypeptides that were not crosslinked appeared along a diagonal since they had the same mobility in both dimensions. Crosslinked proteins appeared below the diagonal, with the partners in a crosslinked oligomer running along the same vertical line. When nuclei isolated from undifferentiated or differentiated Friend cells were oxidized with H_2O_2 and digested with micrococcal nuclease at 10°C

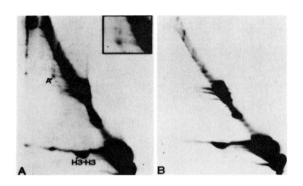

Fig. 1. Two-dimensional gel electrophoresis of extracts from micrococcal nuclease-digested nuclei. (A) Extract from nuclei from undifferentiated Friend cells. (B) Extract from nuclei from differentiated Friend cells. Nuclei were prepared, treated with H_2O_2 and digested with micrococcal nuclease as explained (Grebanier, Pogo 1981). The digested nuclei were incubated with 1 mM EDTA, 0.5 mM phenylmethylsulfonyl fluoride (pH 7.0) for 30 min in an ice bath, then centrifuged for 10 min at 500 x g. The supernatants were concentrated, heated at 90°C for 2 min in the presence of sodium dodecyl sulfate, and run in two-dimensional gels as described (Grebanier, Pogo 1981). The samples were treated with 2-mercaptoethanol between the first and second dimension. Notice the absence of non-histone protein A crosslinked to H3 in extract obtained from differentiated erythroleukemia cells. (From Grebanier, Pogo 1981). <u>Inset</u> is a larger print of area of electrophoretogram with off-diagonal spots.

for 45 min, 6 and 10% of the DNA were rendered soluble, respectively. The nuclei were subsequently extracted with 1 mM EDTA to release the nucleosomes, and analyzed for crosslinked proteins. The histone H3 homodimer was the most abundant species detected (Fig. 1). In samples extracted from nuclei from undifferentiated cells, the most prominent nonhistone protein observed to be involved in crosslinking was a protein of M_r 46,000 (A in Fig. 1A). The off-diagonal spot of protein A is vertically aligned with an off-diagonal spot of histone H3, and the mobility

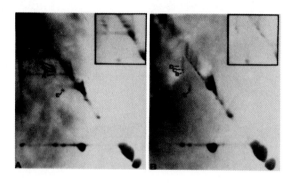

Fig. 2. Two-dimensional gel electrophoresis of high-salt extract from DNAase I-digested nuclei. (A) Extract from nuclei from undifferentiated Friend cells. (B) Extract from nuclei from differentiated cells. Nuclei were isolated, treated with H_2O_2 and digested with DNAase I as explained (Grebanier, Pogo 1979, 1981). After the digestion, the suspension was layered over 5 volumes of 300 mM KCl, 10 mM Tris-HCl (pH 7.7 at 23°C), 1.5 mM $MgCl_2$, 0.5 phenylmethylsulfonyl fluoride and 146 mM sucrose, and the nuclei collected by centrifugation for 10 min at 500 x g. The nuclei were resuspended with 1 M KCl in the same buffer and kept for 30 min in an ice bath; finally they were centrifuged for 10 min at 500 x g. The high-salt extracts were concentrated and analyzed by two-dimensional gel electrophoresis as is described in Figure 1. (From Grebanier, Pogo 1981.) Insets are larger prints of area electrophoretogram with off-diagonal spots.

on the first dimension of the oligomer from which these spots are derived is appropriate for a protein with the size of a complex protein A and histone H3 (Grebanier, Pogo 1979, 1981). Thus, the formation of a disulfide bond between these proteins is proven in undifferentiated cells. In contrast to the results with undifferentiated cells, crosslinked, nonhistone proteins were not seen in EDTA extracts of micrococcal nuclease-digested nuclei from differentiated cells (Fig. 1B).

Following extensive digestion of nuclear DNA with DNAase I, the intranucleosomal histones are extracted coordinately into a buffer containing 1 M KCl (Long et al. 1979). The most obvious crosslinked, nonhistone proteins in the high-salt extract of nuclei from undifferentiated cells were three proteins of $M_r \sim 50,000$ (B, C, D in Fig. 2A) and a protein of $M_r \sim 36,000$ (E in Fig. 2A). Protein A was only marginally detectable in the off-diagonal region of the two-dimensional electrophoretogram. High-salt extracts of nuclei from differentiated cells were similar to those from undifferentiated cells, except that protein A was not detectable on the two-dimensional electrophoretograms (Fig. 2B).

In summary, the profile of released, crosslinked, nonhistone proteins to histone H3 depends upon the protocol followed to prepare the extract. In extracts that contain nucleosomes, i.e., both the DNA and histones, the major crosslinked nonhistone protein found is protein A. By contrast, crosslinked protein A is only marginally detectable in the off-diagonal region of a two-dimensional electrophoretogram of high-salt extract of DNAase I-treated nuclei; the four prominent crosslinked proteins are B, C, D and E. The differences in the pattern of released nonhistone proteins may be caused by the involvement of different kinds of bonds to stabilize the normal interactions between proteins in the nucleus. The difference in the extractability of A versus B, C, D and E is intriguing and may reflect structural heterogeneity. The proximity of these proteins to histone suggest that they may have important roles since H3 plays an essential function in nucleosomal assembly (Jackson, Chalkley 1981). It is possible that these proteins form a link between the histones and the underlying nuclear structure or nuclear matrix. However, given our present state of knowledge, much work remains to be accomplished before this assumption is proven.

SnRNA CROSSLINKED BY AMT TO HnRNA

SnRNA and hnRNA remain in the residual nuclear structure after extensive digestion of nuclei with DNAase I (Long et al. 1979). A high-salt treatment of nuclei or DNA-depleted nuclei does not remove both RNA species, however, 1% sodium deoxycholate produces the selective release, apparently as a ribonucleoprotein complex of snRNA (unpublished observation). The latter observation is consistent with the concept that snRNA and hnRNA are in different nuclear structures.

Although snRNAs have been known to exist for some time, their functions have remained mysterious. Recently it has been proposed that U1 (see Zieve 1981 for nomenclature) may participate in hnRNA processing. The most compelling evidence of this role was obtained when specific antibodies (anti-snRNP) of lupus erythematosus patients were added to an in vitro system containing nuclei from HeLa cells infected with adenovirus, and the selective inhibition of the processing of adenovirus primary transcripts was observed (Lerner, Steitz 1981).

A large homology among U1, U4 and U5 was observed by sequence examination, suggesting a common ancestor (Krol et al. 1981). However, no large homologies were observed for species 4.5, U6, U2 and U3, suggesting that they form another class of snRNA, perhaps having different functions. Notwithstanding, ribonucleoproteins containing U2, U4, U5 and U6 have a common antigenic determinant to U1 snRNA (Lerner, Steitz 1981). In the particular case of U1, formed by variants U1a and U1b which differ only in the middle but are identical at both termini, there is a tetranucleotide A-C-C-Up adjacent to to the cap structure which is complementary to the extremeties (U-G, G-A) of most of the pre-messenger RNA introns studied (Branlant et al. 1981). The existence of unstable U1 snRNA-hnRNA hybrids can be predicted from the pairing of these eight bases in which two A-U pairs will establish weak hydrogen bonds.

Fig. 3 (Next page). Formamide-sucrose gradient centrifugation of crosslinked hnRNA obtained from DNA-depleted nuclei from undifferentiated Friend erythroleukemia cells. The cells were labeled for 24 hr with tritiated uridine. Nuclei were isolated and digested with DNase I as explained (Long

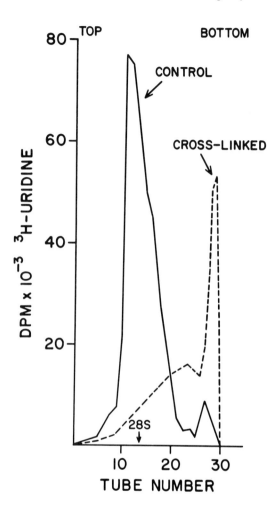

et al. 1979). The DNA-depleted nuclei were resuspended in 10 mM Tris-HCl (pH 7.7 at 23°C), 10 mM EDTA, 1 mM EGTA, and 54.0 µg/ml of AMT was added to one aliquot. The samples were exposed to long wavelength UV light (352 nm) in a Rayonet Type RS photochemical reactor (Southern New England Ultraviolet Co.) for 1 hr, and the temperature was maintained at 4°C with a circulating water bath. The samples were then digested with proteinase K, lysed with lithium dodecyl sulfate and centrifuged in a 10-30% sucrose gradient having a 75% sucrose cushion in the SW41 Spinco rotor at 24,000 rpm for 18 hr. Most of the hnRNA molecules sedimented on top of the 75% sucrose cushion free from rRNA and/or pre-rRNA contaminants (Pogo 1981). Both control and AMT crosslinked hnRNA were alcohol precipitated, resuspended in 50% deionized formamide, and heated at 55°C for 2.5 min. They were then cooled, layered in a discontinuous formamide-sucrose gradient (8-20%) and centrifuged in the SW41 Spinco rotor at 41,000 rpm for 28 hr. As a marker, 28S RNA was denatured and centrifuged under the same conditions. The gradients were fractionated in an ISCO density gradient fractionator. (——) control sample; (---) crosslinked sample. The direction of sedimentation is from left to right.

It is important to obtain such hybrids in order to establish specifically which snRNA interacts with most of the hnRNA, and which bases are involved in these interactions. A covalent linkage was formed between these two RNAs when DNA-depleted nuclei were irradiated with UV light (352 nm) in the presence of AMT. The advantage of using DNA-depleted nuclei in the crosslinking reaction is that there is little DNA and no Mg^{2+} present in the buffer solution. It is known that in the presence of Mg^{2+}, AMT crosslinks form inefficiently, and in whole nuclei, very little AMT binds to RNA (Frederiksen, Hearst 1979).

In a typical protocol, cells were labeled for 24 hr with [^{3}H]uridine; DNA-depleted nuclei were obtained as explained (Long et al. 1979) and AMT was added to one aliquot and then UV irradiated (352 nm). Both nontreated and treated samples were then digested with proteinase K, and finally lysed with lithium dodecyl sulfate. The lysate was fractionated in a sucrose gradient; RNA molecules with a sedimentation coefficient greater than 45S were collected and heated at 65°C in 50% formamide and re-centrifuged in a formamide-sucrose gradient (Pogo 1981). As expected, most of the crosslinked molecules cannot be denatured under conditions which produce complete denaturation of the untreated sample (Fig. 3). In the crosslinked sample, top, middle, and bottom regions of the formamide-sucrose gradient were electrophoresed in polyacryalamide gels; one aliquot was photochemically reversed with UV light (250 nm) and the other was used as a control. The most promiment photoreversal products were polynucleotides with electrophoretic mobilities similar to U1b and U6. Faint bands with mobilities similar to U2, U3 and U5 were detected, and none were observed in the non-photoreversed sample (Fig. 4). The presence of U3 is due to some contamination with 45S pre-rRNA since this snRNA is confined to the nucleolus. None of these bands were observed in the photoreversed but not cross linked samples, indicating that UV photo reversal per se cannot produce snRNA (unpublished observation).

The photoreversed products moved somewhat more slowly than the control, and this was attributed to the presence of AMT monoadducts (Rabin, Crothers 1979). Although similar mobilities do not establish chemical identity, the experimental evidence suggest that U1 and U6 can be effectively crosslinked to hnRNA. The U1 is the snRNA most abundantly crosslinked to hnRNA, indicating that most of the hnRNA must have a common sequence for U1 interaction. It remains to be established whether the site of interaction is the exon-intron junction, and whether the tetranucleotide A-C-C-Up near the 5' end of the U1 is involved in the crosslinking reaction.

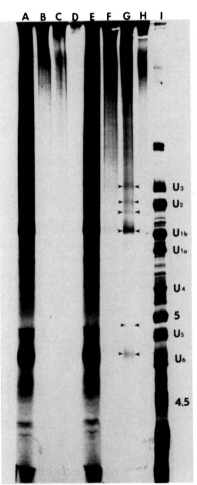

Fig. 4. Analysis of snRNA species cross linked to hnRNA. The crosslinked hnRNA were obtained as is explained in Figure 3. Top, middle, bottom and pellet fractions of the formamide-sucrose gradients were made 0.1 M with LiCl, and ethanol-precipitated in the cold. They were then resuspended in distilled water, and one aliquot was photoreversed with short UV light (252 nm) as is explained (Rabin, Crothers 1979). A 10% acrylamide slab gel was cast containing 7 M urea and overlayed with a narrow, 5% acrylamide spacer gel. Equivalent volumes of the photoreversed and non-photoreversed samples were applied to each well. Samples were heated in 50% formamide in a boiling water bath for 2 min and run at 360V for 25 hr in the Tris-borate-EDTA buffer system of Peacock and Dingman (1967). Gels for fluorography were processed according to Bonner and Laskey (1974). A-D, non-photoreversed samples; E-H, photoreversed samples; I, total snRNA. A and E, top fractions; B and F, middle fractions; C and G, bottom fractions; D and H, D and H, pellet fractions.

In summary, covalent linkages have been formed between a set of snRNA and hnRNa when DNA-depleted nuclei were irradiated with UV light in the presence of AMT. SnRNA species U1 was the most predominantly crosslinked RNA, therefore, most of the hnRNA or pre-mRNA contain a common sequence that regularly interacts with this snRNA. This is the first time that snRNA-hnRNA hybrids have been obtained.

ACKNOWLEDGMENTS

This work was supported by grants from the National Science Foundation and National Institutes of Health. We gratefully acknowledge the secretarial assistance provided by Paula Krauss. Dr. L. Cornudella was supported by a grant from the J. March Foundation at the time of this research.

Bonner WM, Laskey RA (1974). A film detection method for tritium-labelled proteins and nucleic acids in polyacrylamide gels. Eur J Biochem 46:83.
Branlant C, Krol A, Ebel J-P, Gallinaro H, Lazar E, Jacob M (1981). The conformation of chicken, rat and human U1A RNAs in solution. Nucl Acids Res 9:841.
Frederiksen S, Hearst JE (1979). Binding of 4'-aminomethyl 4,5'8-trimethyl psoralen to DNA, RNA and protein in HeLa cells and Drosophila cells. Biochim Biophys Acta 563:343.
Grebanier AE, Pogo AO (1979). Cross-linking of proteins in nuclei and DNA-depleted nuclei from Friend erythroleukemia cells. Cell 18:1091.
Grebanier AE, Pogo AO (1981). Nonhistone proteins crosslinked by disulfide bonds to histone H3 in nuclei from Friend erythroleukemia cells. Biochemistry 20:1094.
Jackson V, Chalkley R (1981). A reevaluation of new histone deposition on replicating chromatin. J Biol Chem 256:5095.
Kornberg RD (1974). Chromatin structure: a repeating unit of histones and DNA. Science 184:868.
Krol A, Branlant C, Lazar E, Gallinaro H, Jacob M (1981). Primary and secondary structures of chicken, rat and man U4 RNAs. Homologies with U1 and U5 RNAs. Nucl Acids Res 9:2699.
Lerner RM, Steitz JA (1981). Snurps and Scyrps. Cell 25:298.
Long BH, Huang C-Y, Pogo AO (1979). Isolation and characterization of the nuclear matrix in Friend erythroleuke-

mia cells: chromatin and hnRNA interactions with the nuclear matrix. Cell 18:1079.

Peacock AC, Dingman CW (1967). Resolution of multiple ribonucleic acid species by polyacrylamide gel electrophoresis. Biochemistry 6:1818.

Pogo AO (1981). Heterogeneous nuclear RNA-protein complexes and nuclear matrix. In Busch H (ed): "The Cell Nucleus", Vol 8, New York: Academic Press, p 331.

Rabin D, Crothers DM (1979). Analysis of RNA secondary structure by photochemical reversal of psoralen crosslinks. Nucl Acids Res 7:689.

van Holde KE, Sahasrabuddhe CG, Shaw BR (1974). A model for particulate structure in chromatin. Nucl Acids Res 1:1579

Wang K, Richards RM (1974). An approach to nearest neighbor analysis of membrane proteins. J Biol Chem 249:8005.

Zieve GW (1981). Two groups of small stable RNAs. Cell 25:296.

ON THE BINDING OF HOST AND VIRAL RNA TO A NUCLEAR MATRIX

Walther J. van Venrooij, Chris A.G. van Eekelen, Edwin C.M. Mariman and Rita J. Reinders

Department of Biochemistry, University of Nijmegen, Geert Grooteplein Noord 21, 6525 EZ Nijmegen

During transcription the nascent hnRNA chains are complexed with proteins (Beyer et al. 1980; for a review see van Venrooij, Janssen 1978). As a result of this association with protein, the long hnRNA chains are foreshortened considerably and it seems that in this way the cell can handle (by condensing) and protect (by covering it with proteins) long, unstable lengths of RNA while still leaving them available for further enzymatic modification. The major protein components of hnRNP particles from HeLa cells can be separated into three groups of closely spaced doublets; the A group (32,000 and 34,000 daltons), the B group (36,000 and 37,000 daltons) and the C group (42,000 and 44,000 daltons). Protein and salt dissociation studies have revealed that the C group proteins are most tightly associated with the hnRNA (Beyer et al. 1977).

Several studies have shown that hnRNA is bound to a higher ordered proteinaceous structure, usually referred to as nuclear matrix (Herman et al. 1978; Miller et al. 1978; van Eekelen, van Venrooij 1981). It is probable that this intranuclear structure plays an important role in the processing and transport of transcripts to the cytoplasm. This article summarizes briefly the evidence obtained in our laboratory that both host- and virus-specific hnRNA molecules are bound to the nuclear matrix via the C group proteins and that this association is maintained during processing of the transcripts.

ISOLATION OF NUCLEAR MATRICES WITH ASSOCIATED hnRNA

Our method of nuclear matrix isolation has been described in detail by van Eekelen and van Venrooij (1981). The critical steps in the isolation of nuclear matrices containing high molecular weight RNA are the digestion of the DNA and the subsequent removal of the chromatin fragments. It is our experience (see also Berezney 1979) that the continuous presence of an inhibitor of protease activities (for example 0.5 mM phenylmethylsulfonyl chloride) is an absolute requirement for the isolation not only of intact matrices but also of high molecular weight matrix RNA. Even the best commercial DNase 1 preparations contain traces of RNase and protease activities which have to be removed or inactivated. Furthermore, the inevitable high salt extraction [we use 0.4 M $(NH_4)_2SO_4$ which is somewhat less rigorous than 2 M NaCl] should preferably be preceded by a centrifugation of the DNase-treated nuclei over a sucrose layer to remove most of the chromatin fragments. Otherwise uncoiling of some remaining, protein-protected chromatin fragments during the high salt treatment may cause mechanical disruption of the fragile internal structure. The whole procedure, starting from intact cells, takes about 90 min and reproducibly results in matrix preparations as shown in Figure 1A. When the same procedure was applied to nuclei isolated from HeLa cells 18 hr after infection with adenovirus type 2, nuclear matrix structures as shown in Figure 1B were obtained. The matrix preparations generally contain less than 1% of the cellular DNA and more than 95% of the rapidly labeled viral and host hnRNA.

In a recent paper of Kaufmann et al. (1981) it was suggested that disulfide bonds between nuclear proteins formed during fractionation could play a role in the formation of these intranuclear structures. Even so, RNase treatment before the high salt wash was shown to be disruptive for the nuclear matrix structure. We have prepared nuclear matrices in which the RNA was exhaustively digested before the high salt treatment and in the continuous presence of 10 mM of the sulfhydryl blocking agent, iodoacetamide, or in the presence of 1 mM 2-mercaptoethanol. Light microscopic observation showed that they had a structure very similar to the ones prepared without these extra additives (compare Fig. 1). In contrast to the results of Kaufmann et al. (1981), our residual nuclear structures all contained prominent nucleoli implying that they were not

Binding RNA to Nuclear Matrix / 237

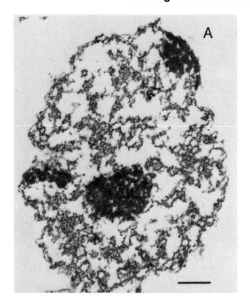

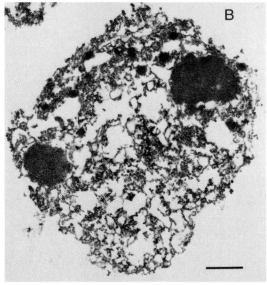

Fig. 1 Nuclear matrices of uninfected and adenovirus-infected HeLa cells. Nuclear matrices from uninfected cells (1A) and from cells 18 hr after infection (1B) were prepared for electron microscopy as described earlier (van Eekelen, van Venrooij 1981). Bars 1 μm.

empty lamina. These results suggest that artificial disulfide crosslinking during cell fractionation, at least in our method of preparation and with the type of cell we use, does not play an essential role in obtaining nuclear matrix structures. Extensive treatment of the matrices (isolated in the presence of SH blocking agents or not) with high concentrations of RNase A after high salt extraction did remove more than 95% of the RNA but did also not significantly alter the structural appearance of the matrices (van Eekelen et al. to be published; van Eekelen, van Venrooij 1981). The remaining RNA fragments (about 4% of the total nuclear matrix RNA) were relatively small (1-50 nucleotides) and probably protected against degradation by their packing with proteins. It seems unlikely that a small amount of such short RNA fragments plays an essential role in the structural stability of the nuclear matrix.

From the results described in this paragraph we concluded that rapidly labeled hnRNA is associated with an internal nuclear proteinaceous structure.

PROTEINS INVOLVED IN THE ASSOCIATION OF HeLa hnRNA WITH THE NUCLEAR MATRIX

Although numerous workers have identified proteins which were present in isolated hnRNP particles, these studies have the generally acknowledged disadvantage that non-specific binding of proteins to the RNA during cell fractionation might have occurred. Furthermore, no definite information is obtained about the proteins that directly interact with the RNA as some proteins may be associated with the RNP particles, for example, via protein-protein interactions. To identify the proteins that are tightly associated with hnRNA we have used irradiation with ultraviolet (UV) light (254 nm) as a method for RNA-protein crosslinking (Wagenmakers et al. 1980; Greenberg 1980; van Eekelen, van Venrooij 1981). To be certain that only proteins interacting in vivo with RNA were covalently linked, we performed the irradiation on intact cells. Only proteins closely associated with RNA can be covalently linked by UV irradiation (for references see Wagenmakers et al. 1980) and isolation of the covalent RNA-protein complexes can be performed under conditions that exclude co-purification of non-specifically bound proteins. Furthermore, by labeling of the RNA moiety of the RNA-protein complexes, the proteins covalently linked with the RNA can specifically be identified after nuclease digestion and subsequent gel electrophoresis by fluorography, owing to the residual radioactive nucleotides bound to them (van Venrooij et al. 1981; van Eekelen et al. 1981a). Using this method we were able to show that HeLa cell hnRNA in vivo is most tightly associated with proteins of 41,500 and 43,000 molecular weight (van Eekelen, van Venrooij 1981). These proteins comigrate on sodium dodecyl sulfate (SDS)-polyacrylamide gels with the C group proteins described by Beyer et al. (1977). Longer exposure times of the gels revealed that also the B group proteins were crosslinked to rapidly labeled hnRNA, albeit less efficiently (van Eekelen et al. 1981a, b). The A group proteins could apparently not be crosslinked with RNA (see Fig. 2). It should be mentioned here that the in vivo crosslinking pattern was identical to the crosslinking pattern obtained in vitro, that is after irradiation of isolated nuclear matrices (van Eekelen, van Venrooij 1981). We have also shown that the proteins crosslinked in vivo to hnRNA are different from those crosslinked in vivo to cytoplasmic mRNA (van Eekelen et al. 1981a).

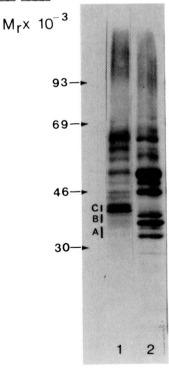

Fig. 2. HnRNP proteins directly and indirectly associated with the nuclear matrix of HeLa cells. Matrix-associated proteins were labeled with [^{35}S]methionine as described earlier (van Eekelen, van Venrooij 1981). Lane 1 shows the proteins crosslinked to poly(A)-containing hnRNA by UV irradiation of intact cells. Lane 2 shows the proteins released from the nuclear matrix by RNase A digestion. For details see van Eekelen, van Venrooij (1981).

Having identified some proteins which in vivo are associated with hnRNA, we attempted to distinguish between matrix proteins involved in the binding of hnRNA and packaging proteins associated with the hnRNA but not with the matrix. Since we had found that the matrix itself was ribonuclease-resistant we expected that a ribonuclease treatment would only release proteins associated with the matrix via RNA. Indeed, treatment of nuclear matrices with high concentrations of RNase A released more than 95% of the matrix-associated RNA but only 1-2% of the matrix proteins (van Eekelen, van Venrooij 1981). Three to five of the released proteins were found in the 33,000-38,000 molecular weight range and comigrated on SDS-polyacrylamide gels with the group A and B proteins described by Beyer and coworkers (1977) as being the main hnRNP subparticle proteins (Fig. 2, lane 2). The C group proteins were not released by the ribonuclease although the crosslinking experiment had shown that these proteins are tightly bound with hnRNA in vivo as well as in vitro. From these results we concluded that hnRNA is bound to the nuclear matrix via the C group proteins. The B group proteins evidently are in direct contact with the hnRNA but not with the matrix structure.

HOST PROTEINS ARE INVOLVED IN THE BINDING OF ADENOVIRUS-SPECIFIC hnRNA TO THE MATRIX

Using the same methodology described above we have also investigated whether hnRNA in virus-infected cells was attached to the nuclear matrix and, if so, if it was associated with the same set of proteins as found associated with host hnRNA. For this purpose, we used HeLa S3 cells infected for 18 hr with adenovirus type 2. The gradient profiles of rapidly labeled hnRNA extracted from nuclei and nuclear matrices from infected cells show a close resemblance. About 95% of the total nuclear RNA from infected cells was found to be associated with matrix structures. We have verified the presence of virus-specific sequences in nuclear matrix RNA by hybridization with adeno DNA fragments. Figure 3 shows the profile of matrix RNA containing nucleotide sequences complementary to the DNA strands of the EcoR1 restriction fragment B (58.5-70.7 m.u.).

In agreement with the results of Beltz and Flint (1979), we found that the hnRNA from nuclei or nuclear matrices of cells late after infection with adenovirus is only partly virus-specific (van Eekelen et al. 1981b). To study the adenovirus-specific hnRNA-protein complexes, crosslinked in vivo by UV irradiation, we separated viral from host hnRNA-protein complexes by hybridization of the RNA moiety of the complex to adenoviral DNA coupled to Sepharose. The hybridization conditions (presence of 0.5% SDS, 0.75 M NaCl and 50% formamide; see van Eekelen et al. 1981b for details) do not allow native uncrosslinked hnRNP to be hybridized to the adeno DNA-Sepharose without complete hnRNP disaggregation. Subsequent analysis of the viral hnRNP complexes showed that adenoviral hnRNA was crosslinked in vivo to proteins of 41,500 and 43,000 molecular weight. These proteins exactly comigrate with the major proteins crosslinked to hnRNA from uninfected cells

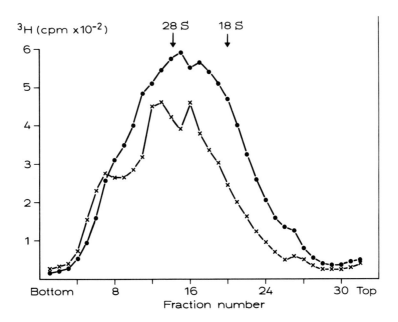

Fig. 3. Sucrose gradient profile of nuclear matrix hnRNA from HeLa cells late after infection with adenovirus. Labeling and isolation of rapidly labeled hnRNA from nuclear matrices was performed as described by Mariman et al. (1981). o-o-o, total nuclear matrix RNA; x-x-x, matrix RNA complementary to the EcoR1 B fragment of adenovirus DNA. For details see Mariman et al. (1981).

(compare Fig. 4, lanes 2 and 3), indicating that adenoviral transcripts are associated with the nuclear matrix via these host proteins.

HnRNA IS ASSOCIATED WITH THE NUCLEAR MATRIX DURING SPLICING

After its synthesis virus-specific hnRNA molecules are capped, methylated, polyadenylated and spliced in much the same way as the hnRNA of eukaryotic cells (Darnell 1979; Ziff 1980). It would be interesting to know if these processing steps occur while the RNA molecules are bound to the matrix. In particular, we attempted to answer the question

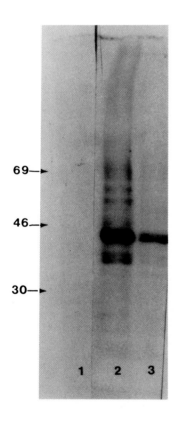

Fig. 4. Analysis of proteins crosslinked in vivo to host and adenovirus-specific hnRNA. Uninfected and infected HeLa cells were labeled with [^{3}H]uridine and [^{3}H]cytidine and one portion of the cells was irradiated with UV light as described (van Eekelen, van Venrooij 1981). From the unirradiated and irradiated uninfected cells nuclear matrix RNA was prepared and hnRNP complexes were selected over oligo-(dT)-cellulose. The adenovirus-specific hnRNP complexes were isolated via adeno DNA-Sepharose columns (van Eekelen et al. 1981b). The purified [^{3}H]-labeled hnRNP particles were extensively nuclease treated (van Eekelen, van Venrooij 1981) and analyzed on 10-18% SDS-polyacrylamide gradient gels. The crosslinked proteins were visualized by fluorography.
Lane 1: hnRNP complexes from unirradiated cells.
Lane 2: hnRNP complexes from irradiated uninfected cells.
Lane 3: Adenovirus-specific hnRNP complexes from irradiated infected cells.

whether splicing of pre-mRNA molecules is a matrix-associated process. For this purpose the matrix-bound poly(A)-containing RNA from adenovirus-infected cells was analyzed by an S1 mapping procedure, described in detail elsewhere (Mariman et al. 1981). When the EcoR1 B fragment of adenovirus DNA was used for the hybridization, seven

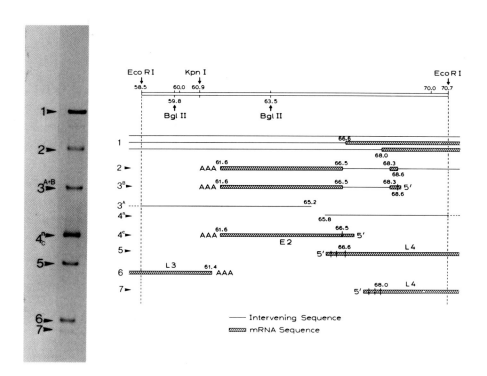

Fig. 5. Pattern of poly(A)-containing RNA sequences present in nuclear matrices. [^{3}H]uridine-labeled matrix RNA from adenovirus-infected HeLa cells was prepared by phenol extraction and selected over oligo(dT)-cellulose. The poly(A)-containing RNA was hybridized to the EcoR1 B fragment of adenovirus DNA and analyzed via an S1 mapping procedure (Mariman et al. 1981). The gel lane at the left shows an autoradiograph of the S1-resistant hybrids. The numbers refer to the right part of the figure which summarizes the most probable identities of these hybrid bands. For details, see Mariman et al. (1981).

major types of RNA molecules could be distinguished (Fig. 5). Four of them, bands number 4^c, 5, 6 and 7 were identified as processed mRNA molecules (Mariman et al. 1981). Bands 1 and 2 correspond to precursors of the region L4 and L5 mRNAs and to the precursor of the DNA binding protein (DBP) mRNA, respectively. Bands 3 and 4^n were tentatively identified as processing intermediates generated from the DBP and L4-L5 region. The implication of these results, which have been described in detail elsewhere (Mariman et al. 1981), is that pre-mRNA is associated with the nuclear matrix during splicing.

Our working hypothesis is that all components necessary for the splicing process are assembled in a splicing complex localized on the matrix structure. In agreement with such an hypothesis is the finding that precursors, intermediates and products of RNA processing can be crosslinked in vivo, by means of irradiation with UV light, to host proteins of 41,500 and 43,000 molecular weight (Mariman et al. 1981). These same proteins are involved in the binding of host hnRNA to the matrix (see above). The association of the various nuclear RNA species (precursors, splicing intermediates and products) with these matrix proteins evidently is not influenced by the processing of the pre-mRNAs to mRNA. One of the questions being studied now at our laboratory is the possibility that specific RNA sequences are involved in the association of hnRNA to the nuclear matrix.

ACKNOWLEDGMENTS

This work was supported, in part, by the Netherlands Foundation for Chemical Research (SON) and the Netherlands Organization for the Advancement of Pure Research (ZWO).

Beltz GA, Flint SJ (1979). Inhibition of HeLa cell protein synthesis during adenovirus infection. J Mol Biol 131: 353.
Beyer AL, Christensen ME, Walker BW, Le Stourgeon WM (1977). Identification and characterization of the packaging proteins of core 40S hnRNP particles. Cell 11:127.
Beyer AL, Miller OL Jr, McKnight SL (1980). Ribonucleoprotein structure in nascent hnRNA is nonrandom and sequence-dependent. Cell 20:75.

Berezney R (1979). Effect of protease inhibitors on matrix proteins and the association of replicating DNA. Exp Cell Res 123:411.
Darnell JE (1979). Transcription units for mRNA production in eukaryotic cells and their DNA viruses. Prog Nucl Acid Res Mol Biol 22:327.
van Eekelen CAG, van Venrooij WJ (1981). HnRNA and its attachment to a nuclear protein matrix. J Cell Biol 88:554.
van Eekelen CAG, Reimen T, van Venrooij WJ (1981a). Specificity in the interaction of hnRNA and mRNA with proteins as revealed by in vivo cross-linking. FEBS Lett 130:5685.
van Eekelen CAG, Mariman ECM, Reinders RJ, van Venrooij WJ (1981b). Adenoviral hnRNA is associated with host cell proteins. Eur J Biochem, in press.
Greenberg JR (1980). Proteins cross-linked to mRNA by irradiating polyribosomes with ultraviolet light. Nucl Acids Res 8:5685.
Herman R, Wymouth L, Penman S (1978). Heterogeneous nuclear RNA-protein fibers in chromatin-depleted nuclei. J Cell Biol 78:663.
Kaufmann SH, Coffey DS, Shaper JH (1981). Considerations in the isolation of rat liver nuclear matrix, nuclear envelope and pore complex lamina. Exp Cell Res 132:105.
Mariman ECM, van Eekelen CAG, Reinders RJ, Berns AJM, van Venrooij WJ (1981). Adenoviral heterogeneous nuclear RNA is associated with the host nuclear matrix during splicing. J Mol Biol in press.
Miller TE, Huang CY, Pogo AO (1978). Rat liver nuclear skeleton and RNP complexes containing hnRNA. J Cell Biol 76:675.
van Venrooij WJ, Janssen DB (1978). HnRNP particles. Mol Biol Rep 4:13.
van Venrooij WJ, Wagenmakers AJM, van den Oetelaar P, Reinders RJ (1981). In vivo cross-linking of proteins to mRNA in human cells. Mol Biol Rep 7:93.
Wagenmakers AJM, Reinders RJ, van Venrooij WJ (1980). Cross-linking of mRNA to proteins by irradiation of intact cells with ultraviolet light. Eur J Biochem 112:323.
Ziff EB (1980). Transcription and RNA processing by the DNA tumor viruses. Nature 287:491.

THE NUCLEAR MATRIX IN STEROID HORMONE ACTION

Evelyn R. Barrack, Ph.D.

Department of Urology
The Johns Hopkins University School of Medicine
Baltimore, Maryland 21205

The interaction of steroid hormones and their specific receptor proteins with the nucleus of target tissues is an essential step in the mechanism by which these hormones modulate nuclear events such as gene expression (see reviews by Thrall et al. 1978; Jensen 1979; Liao et al. 1979). The hormone-receptor complex is presumed to bind to specific regulatory sites near the genes they regulate, but the identification of these binding sites and the precise mechanism of transcriptional regulation is still unresolved. Evidence is accumulating to suggest that the nuclear matrix may play a fundamental role in many dynamic aspects of nuclear function, including DNA replication and heterogeneous nuclear RNA synthesis (see reviews by Berezney, Coffey 1976; Shaper et al. 1979; Agutter, Richardson 1980; Barrack, Coffey 1982). Since steroid hormones affect these processes in specific target tissues, it is of interest that hormones interact in a specific manner with the nuclear matrix (Barrack et al. 1977, 1979; Barrack, Coffey 1980, 1982).

We have identified and characterized specific sex steroid binding sites associated with the nuclear matrix of both estrogen and androgen responsive tissues. The properties of these specific binding sites for estradiol on the nuclear matrix of rat uterus and chicken liver, and for dihydrotestosterone (DHT) on the nuclear matrix of rat ventral prostate, satisfy the criteria that are commonly used to characterize steroid hormone receptors. The binding of these steroids is saturable, high affinity ($K_d \sim 1$ nM), and heat and pronase sensitive. In addition, the steroid specificity of these sites is characteristic of estrogen and androgen receptors, respectively. The specific binding of [^{3}H]estradiol to the nuclear matrix of estrogen target tissues is inhibited by unlabeled estrogens, but not by androgens, progestins or cortisol. Similarly, the binding of [^{3}H]DHT to the nuclear matrix of androgen target tissues is inhibited by androgens, but 100-1000 fold less effectively by other classes of steroids (Barrack, Coffey 1980).

In these tissues, the appearance of specific steroid binding sites with the nuclear matrix occurs in response to an appropriate hormonal stimulus, not indiscriminately. Thus, for example, the liver of egg-laying hens, in response to high blood levels of estrogen, synthesizes vitellogenin (the precursor of the major egg yolk proteins); the liver nuclear matrix of hens binds about 42 fmol estradiol/100 μg starting nuclear DNA equivalents or 600 molecules/nucleus equivalent. In contrast, the liver nuclear matrix of untreated roosters, which do not produce yolk proteins, contains only one-eighth as many specific binding sites for estradiol (5.5 fmol/100 μg nuclear DNA equivalents) as that of the laying hen (Fig. 1). However, the administration of pharmacological doses of estrogen (25 mg/kg) to roosters or immature chicks results in a stimulation of vitellogenin mRNA synthesis (Deeley et al. 1977) and a marked increase (12-fold) in the number of nuclear matrix-associated specific estradiol binding sites (65 fmol/100 μg nuclear DNA equivalents; Fig. 2).

Similarly, in the ventral prostate the presence of specific DHT binding sites on the nuclear matrix is associated with androgen stimulation of this gland. These binding sites are present in the prostate nuclear matrix of intact adult male rats; but following withdrawal of androgen (castration), there is a rapid loss (within 24 hr) of these nuclear matrix binding sites that precedes the involution of the gland (Fig. 3). Administration of androgens, but not of estrogens, restores these sites to normal levels within 1 hr (Barrack, Coffey 1980). In order to quantitate androgen binding sites on the prostate nuclear matrix, the inclusion of phenylmethylsulfonyl fluoride in the isolation buffers is essential. The degradation of binding sites and decreased recoveries of nuclear matrix spheres and protein in the absence of protease inhibitor are probably due to the high concentration of hydrolytic and proteolytic enzymes normally present in the prostate.

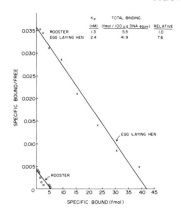 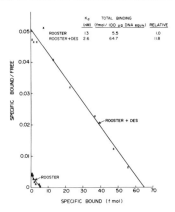

Fig. 1. (left) Scatchard plots of specific binding of estradiol to liver nuclear matrix isolated from hens and roosters. Binding assays were carried out (4°C, 24 hr) in the presence of 0.15 to 20 nM [^{3}H]estradiol alone (to measure total binding) or together with 1 µM unlabeled estradiol (to measure nonspecific binding); free steroid was removed by washing. Specific binding was calculated by subtracting nonspecific from total binding.

Fig. 2. (right) Diethylstilbestrol (DES) treatment of roosters increases specific estradiol binding sites on liver nuclear matrix.

Following treatment of the immature female rat with a physiological dose of estrogen (0.1 µg) that is sufficient to induce maximal uterine growth, specific estradiol binding sites are found associated with the uterine nuclear matrix, but not with the liver nuclear matrix of these animals (Barrack et al. 1977). This latter observation is consistent with the lack of responsiveness of the rat liver to these small doses of estrogen (Aten et al. 1978). Recently, Agutter and Birchall (1979) have confirmed that the rat uterine nuclear matrix contains specific estradiol binding sites and have shown in addition that the rat lung nuclear matrix contains none. From all of the above considerations, therefore, we conclude that the appearance of nuclear matrix-associated steroid binding sites correlates with the stimulation of a biological response.

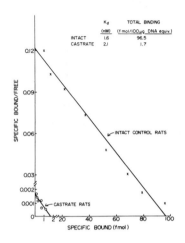

Fig. 3. Scatchard plots of specific binding of DHT to prostate nuclear matrix of intact and castrate rats.

Clark and his colleagues have recently demonstrated that rat uterine nuclei contain, in addition to the well-known high affinity (Type I, $K_d \sim 1$ nM) estradiol receptors, low affinity (Type II, $K_d \sim 30$ nM) specific estradiol binding sites (Clark et al. 1980). We have looked for low affinity, Type II binding of steroid to the nuclear matrix of avian liver and rat prostate by carrying out saturation analyses in the presence of concentrations of labeled steroid ranging from 0.1 nM to 42 or 80 nM, but find only a single class of specific, high affinity ($K_d \sim 1$ nM) steroid binding sites.

A significant proportion of the steroid binding capacity of nuclei is localized to the nuclear matrix. In the avian liver, for example, the recovery of estradiol binding sites with the nuclear matrix constitutes 61% of the total number of specific estradiol binding sites in unextracted nuclei (Table 1). In contrast, only 7% of the total nuclear protein and 2% of the DNA are recovered with the nuclear matrix; thus, per unit amount of protein, there is a 5-fold enrichment of nuclear estradiol binding sites on the matrix.

Table 1

Comparison of Specific Estradiol Binding in Chick Liver
Nuclei and Nuclear Matrix

	Recovery		Specific binding of estradiol		
	Protein	DNA	Per 100 µg protein in fraction	Per 100 µg starting DNA equivalents	K_d
	% of total nuclear		fmol		nM
Total nuclei	100	100	24.2(1.0)	59.2(1.0)	0.9
Nuclear matrix	7	2	127.3(5.3)	36.2(0.61)	0.5

Liver nuclei and nuclear matrix were isolated from estrogenized chicks. Specific estradiol binding capacities and dissociation constants were derived from Scatchard analyses (see Barrack, Coffey 1980 for experimental details). Numbers in parentheses are values relative to total nuclei.

In the rat prostate (Table 2-A) a similar percentage (50-67%) of the total number of nuclear DHT binding sites is recovered with the nuclear matrix, which contains only 10 to 15% of the total nuclear protein and 1 to 2% of the DNA. The remainder of the nuclear androgen receptors are easily extracted by 2 M NaCl. On a protein basis, the specific activity of the DHT binding sites on the nuclear matrix is 4-fold higher than that of intact nuclei, and 10-fold higher than that of the salt extract (Table 2-A). It is not likely that the nuclear matrix-associated steroid binding sites result from exposure of sites normally masked in the intact nucleus since binding in the soluble extracts and in the insoluble nuclear matrix account for essentially all of the binding that is observed in unextracted nuclei.

The nuclear matrix, which we prepare by extracting nuclei with Triton X-100, DNase I, and 2 M NaCl (Barrack,

Coffey 1980), consists of a peripheral lamina, residual nucleolus, and an internal ribonucleoprotein network. To determine whether the hormone binding sites are distributed uniformly throughout the matrix structure, or are enriched in a specific morphological component of the matrix, we have taken advantage of the observation that when Triton X-100-washed prostate nuclei are treated with DNase I and RNase A in the presence of 1 mM dithiothreitol (DTT), and then extracted with 2 M NaCl, matrix spheres are obtained that are devoid of internal structure, and contain only 60% as much protein as control matrix preparations.

Table 2

Distribution and Enrichment of DHT Binding Sites in Subfractions of Prostate Nuclei and Nuclear Matrix

Pretreatment of nuclei[a]	Distribution			Specific Activity		
	Nuclear matrix	Salt extract	Total	Nuclear matrix	Salt extract	Total
	(fmol DHT/100 µg starting DNA equivalents)			(fmol DHT/100 µg protein)		
A. DNase (control)	101±8	49±14	150±17	194±48	19±6	49±6
B. DNase + RNase+DTT	17±4	136±3	153±2	52±20	44±6	45±3

[a]Pretreatments of the nuclei were followed by extraction with 2 M NaCl to obtain the insoluble nuclear matrix (experiment A) or pore complex-lamina (experiment B), and the soluble salt extract.

The isolation of nuclei and nuclear matrix from prostates of intact rats, and the measurement of specific DHT binding, are described in detail in Barrack, Coffey (1980).

By comparing the distribution and specific activity of the hormone binding sites associated with intact vs. empty matrix spheres, we have been able to show that the matrix-associated steroid binding sites appear to be enriched on internal matrix structures (Barrack, Coffey 1980). Binding sites are not localized exclusively to the lamina, nor are they distributed uniformly on all matrix components. As shown in Table 2-B, the recovery of specific DHT binding sites with empty prostate matrix spheres (consisting only of a peripheral pore complex-lamina) represents only 17% of the DHT binding activity associated with intact nuclear matrix structures that contain internal network material (17 vs. 101 fmol/100 µg starting nuclear DNA equivalents). If the binding sites had been distributed uniformly on all matrix components, then the specific activity of binding in the residual fraction would have been unchanged. We find, however, that the specific activity of the DHT binding sites that remain associated with these empty matrix spheres (52 fmol/100 µg protein) is only 27% of that of intact nuclear matrix structures (194 fmol/100 µg protein). The DHT binding sites that were originally associated with the intact nuclear matrix are recovered in the salt extract, the specific activity of which is thereby increased 2.3 fold (19 vs. 44 fmol/100 µg salt-extractable protein).

The specific activity of these matrix-associated binding sites that become salt-extractable as a result of the additional pretreatment with RNase and DTT (i.e., the difference between the amount of salt-resistant matrix binding and protein in experiments A vs. B) is calculated to be about 400 fmol/100 µg of protein extracted from the matrix; if these DHT binding sites had been associated with the internal ribonucleoprotein network of the nuclear matrix, then the specific activity of these internal sites is almost 8-fold greater than that of the sites remaining with the peripheral lamina fraction (52 fmol/100 µg protein). A similar distribution of matrix-associated estradiol binding sites is observed in hen liver (Barrack, Coffey 1982). In addition, Agutter and Birchall (1979) find that whereas the nuclear matrix of rat uterine nuclei contains estradiol binding sites, the pore complex-lamina fraction, isolated by a different method, contains virtually none of these sites.

Salt extraction of detergent-treated nuclei that had been digested with DNase and RNase, or DNase and DTT, is less effective in removing intranuclear contents and matrix-associated hormone binding sites (Barrack, Coffey 1980, 1982). Nuclease treatments alone solubilize no binding activity. With regard to the observed effects of DTT, DTT treatment alone does not extract steroid binding sites; it only renders them capable of being solubilized by subsequent or simultaneous treatment with salt. Thus, even if DTT is added only to the tissue homogenizing buffer, extraction of the isolated nuclei with NaCl will result in the solubilization of most of the nuclear receptors. Indeed, this may explain why some investigators do not find significant amounts of salt-resistant binding (Traish et al. 1977; Chamness et al. 1978).

It is important to consider the possibility that the conditions which result in solubilization of the internal ribonucleoprotein network may also extract certain components from the peripheral lamina; hence we cannot rule out the possibility that the hormone binding sites associated with the nuclear matrix in fact may also derive from components of the peripheral lamina. However, with regard to the apparent localization of the majority of the matrix-associated hormone binding sites to the internal ribonucleoprotein network, it is interesting to note that Liao et al. (1973) have reported the ability of prostate cytosol DHT-receptor complexes to bind to isolated nuclear RNP particles that were resistant to solubilization from nuclei by 1 M KCl and DNase I but could be released by deoxycholate treatment. That these RNP-associated binding sites (Liao et al. 1973) may have been a component of the internal RNA-protein network that is observed in chromatin- and lipid-depleted nuclei (Herman et al. 1978; Miller et al. 1978; Faiferman, Pogo 1975), is supported by the observation of Miller et al. 1978) that the RNP complexes of this network are highly susceptible to disruption by deoxycholate. In addition, electron autoradiographic studies have demonstrated that the DNA replication sites (Pardoll et al. 1980) and the newly labeled heterogeneous nuclear RNA (Fakan et al. 1976; Herman et al. 1978) are associated with the internal ribonucleoprotein network structure, and indicate the potential biological importance of this component of the nuclear matrix.

The possibility that these matrix-associated steroid binding sites do not reflect a native steroid-receptor-acceptor interaction, but merely result from adventitious adsorption of receptors must always be considered. The adsorption of unoccupied cytosol receptors to the nuclear matrix, however, can be ruled out, since in the prostate of the castrate one finds large amounts of androgen receptor in the cytosol, but virtually none on the matrix. Nevertheless, one must still consider the possibility that during the isolation of nuclei and nuclear matrix some activated receptors might become associated with the nuclear matrix in a nonphysiological interaction that remains resistant to disruption and solubilization by high ionic strength. On the other hand, however, since the precise mechanism by which steroid hormone-receptor complexes act in the nucleus to modulate the synthesis of premessenger heterogeneous nuclear RNA is still unresolved, the possibility that salt-resistant interactions of receptors with the nuclear matrix in fact reflect a meaningful association should not be discounted (see also Barrack, Coffey 1982; Clark, Peck 1976; Honma et al. 1977; Ruh, Baudendistel 1978; Sato et al. 1979).

The role of hormone binding components of chromatin in the mechanism of steroid hormone action has been investigated extensively (see review, Thrall et al. 1978). Unfortunately, however, chromatin is operationally defined and a variety of different methods are used to prepare and subfractionate it. Hence individual preparations may represent functionally variable components of the nucleus. Chromatin preparations often contain fragments of the nuclear envelope (Jackson 1976) and of the nuclear matrix (Berezney 1979). It should not be surprising to find that isolated chromatin would contain components of the nuclear matrix, particularly if chromatin is prepared by simply washing nuclei with a low ionic strength buffer to extract soluble nuclear proteins (Thrall, Spelsberg 1980). In this regard, it is important to note that tissue-specific nuclear acceptor sites for steroid hormone-receptor complexes have been described in chromatin fractions that resist solubilization by both low and high (0.5-2 M NaCl or 4-5 M GuHCl) ionic strength (Klyzsejko-Stefanowicz et al. 1976; Wang 1978; Thrall et al. 1978; Tsai et al. 1980). In addition, salt-resistant nuclear receptors for estradiol have been implicated in the mechanism for the true growth response of the uterus to estrogens (Clark, Peck 1976; Ruh, Baudendistel 1978) and in the induction by estrogens of Leydig cell tumors in mice (Sato et al. 1979).

ACKNOWLEDGMENTS

This work was supported by grant number AM-22000, USPHS, NIAMDD.

Agutter PS, Birchall K (1979). Functional differences between mammalian nuclear protein matrices and pore-complex laminae. Exp Cell Res 124:453.
Agutter PS, Richardson JCW (1980). Nuclear non-chromatin proteinaceous structures: their role in the organization and function of the interphase nucleus. J Cell Sci 44:395.
Aten RF, Weinberger MJ, Eisenfeld AJ (1978). Estrogen receptors in rat liver: translocation to the nucleus in vivo. Endocrinology 102:433.
Barrack ER, Coffey DS (1979). The specific binding of estrogens and androgens to the nuclear matrix of sex hormone responsive tissues. J Biol Chem 255:7265.
Barrack ER, Coffey DS (1982). Biological properties of the nuclear matrix: steroid hormone binding. Recent Prog Horm Res 38:in press.
Barrack ER, Hawkins EF, Allen SL, Hicks LL, Coffey DS (1977). Concepts related to salt resistant estradiol receptors in rat uterine nuclei: nuclear matrix. Biochem Biophys Res Commun 79:829.
Barrack ER, Hawkins EF, Coffey DS (1979). The specific binding of estradiol to the nuclear matrix. In Leavitt WW, Clark JH (eds): "Steroid Hormone Receptor Systems", New York: Plenum Press, p. 47.
Berezney R (1979). Dynamic properties of the nuclear matrix. In Busch H (ed): "The Cell Nucleus", Vol 7, New York: Academic Press, p. 413.
Berezney R, Coffey DS (1976). The nuclear protein matrix: isolation, structure and functions. Advances in Enzyme Regulation 14:63.
Chamness GC, Zava DT, McGuire WL (1978). Methods for assessing the binding of steroid hormones in nuclei. Meth Cell Biol 17:325.
Clark JH, Peck EJ Jr (1976). Nuclear retention of receptor-oestrogen complex and nuclear acceptor sites. Nature 260:635.
Clark JH, Markaverich B, Upchurch S, Eriksson H, Hardin JW, Peck EJ Jr (1980). Heterogeneity of estrogen binding sites: relationship to estrogen receptors and estrogen responses. Recent Prog Horm Res 36:89.

Deeley RG, Gordon JI, Burns ATH, Mullinix KP, Bina-Stein M, Goldberger RF (1977). Primary activation of the vitellogenin gene in the rooster. J Biol Chem 252:8310.

Faiferman I, Pogo AO (1975). Isolation of a nuclear ribonucleoprotein network that contains heterogeneous RNA and is bound to the nuclear envelope. Biochemistry 14:3808.

Fakan S, Puvion E, Spohr G (1976). Localization and characterization of newly synthesized nuclear RNA in isolated rat hepatocytes. Exp Cell Res 99:155.

Herman R, Weymouth L, Penman S (1978). Heterogeneous nuclear RNA-protein fibers in chromatin-depleted nuclei. J Cell Biol 78:663.

Honma Y, Kasukabe T, Okabe J, Hozumi M (1977). Glucocorticoid binding and mechanism of resistance in some clones of mouse myeloid leukemic cells resistant to induction of differentiation by dexamethasone. J Cell Physiol 93:227.

Jackson RC (1976). On the identity of nuclear membrane and non-histone nuclear proteins. Biochemistry 15:5652.

Jensen EV (1979). Interaction of steroid hormones with the nucleus. Pharmacol Rev 30:477.

Klyzsejko-Stefanowicz L, Chiu J-F, Tsai Y-H, Hnilica LS (1976). Acceptor proteins in rat androgenic tissue chromatin. Proc Natl Acad Sci USA 73:1954.

Liao S, Liang T, Tymoczko JL (1973). Ribonucleoprotein binding of steroid-"receptor" complexes. Nature (New Biol) 241:211.

Liao S, Mezzetti G, Chen C (1979). Androgen receptor and early biochemical responses. In Busch H (ed): "The Cell Nucleus", Vol 7, New York: Academic Press, p. 201.

Miller TE, Huang C-Y, Pogo AO (1978). Rat liver nuclear skeleton and ribonucleoprotein complexes containing hnRNA. J Cell Biol 76:675.

Pardoll DM, Vogelstein B, Coffey DS (1980). A fixed site of DNA replication in eucaryotic cells. Cell 19:527.

Ruh TS, Baudendistel LJ (1978). Anti-estrogen modulation of the salt-resistant nuclear estrogen receptor. Endocrinology 102:1838.

Sato B, Spomer W, Huseby RA, Samuels LT (1979). The testicular estrogen receptor system in two strains of mice differing in susceptibility to estrogen-induced Leydig cell tumors. Endocrinology 104:822.

Shaper JH, Pardoll DM, Kaufmann SH, Barrack ER, Vogelstein B, Coffey DS (1979). The relationship of the nuclear matrix to cellular structure and function. Advances in Enzyme Regulation 17:213.

Thrall CL, Spelsberg TC (1980). Factors affecting the binding of chick oviduct progesterone receptor to deoxyribonucleic acid: evidence that deoxyribonucleic acid alone is not the acceptor site. Biochemistry 19:4130.

Thrall CL, Webster RA, Spelsberg TC (1978). Steroid receptor interaction with chromatin. In Busch H (ed): "The Cell Nucleus", Vol 6, New York: Academic Press, p. 461.

Traish AM, Muller RE, Wotiz HH (1977). Binding of estrogen receptor to uterine nuclei. Salt-extractable versus salt-resistant receptor:estrogen complexes. J Biol Chem 252:6823.

Tsai Y-H, Sanborn BM, Steinberger A, Steinberger E (1980). Sertoli cell chromatin acceptor sites for androgen-receptor complexes. J Steroid Biochem 13:711.

Wang TY (1978). The role of nonhistone chromosomal proteins in the interaction of prostate chromatin with androgen: receptor complex. Biochim Biophys Acta 518:81.

HETEROGENEITY OF ESTROGEN-BINDING SITES AND THE NUCLEAR MATRIX

J.H. Clark and B.M. Markaverich

Baylor College of Medicine
Department of Cell Biology
Texas Medical Center
Houston, Texas 77030

INTRODUCTION

Estrogens enter target cells by simple diffusion and interact with macromolecules in the cytoplasm called receptors. Subsequent to this interaction, the receptor-estrogen complex binds to nuclear acceptor sites which results in changes in RNA, DNA and protein synthesis, which ultimately are required for the stimulation of uterine growth (for review see Clark, Peck 1979). We have previously shown that a limited number (~1000-2000 sites/cell) of the total population (~20,000 sites/cell) of receptor-estrogen complexes must be retained in the nucleus for a prolonged period of time in order to stimulate true uterine growth (Anderson et al. 1972, 1973). These appear to be equivalent in number to those sites which are tightly bound in the nucleus and are resistant to 0.4 M KCl-extraction (Clark, Peck 1976). We suggested that the salt-resistant receptor estrogen-complexes may represent those receptors associated with a limited number of high affinity nuclear acceptor sites (Clark, Peck 1976).

Although provocative in theory, the existence of nuclear acceptor sites for the receptor-estrogen complex has not been directly demonstrated and recently our results have been both confirmed and contradicted. Studies by Juliano and Stancel (1976) failed to demonstrate two classes (high and low affinity) of nuclear-receptor-estrogen complexes suggesting a distinction could not be made between non-acceptor- and acceptor-associated estrogen receptors.

Similarly, Traish et al. (1977) suggested that salt-extraction techniques were not specific enough to discriminate between receptor binding to acceptor and non-acceptor sites, and results obtained by salt-extraction may be artifacts due to the gelatinous nature of nuclear pellet as well as to non-specific binding to nuclear matrix. In contrast, Ruh and Baudendistal (1977) have reported findings with salt-extraction which are identical to ours.

The studies by Barrack et al. (1977) suggested the contradictory results obtained by salt-extraction techniques (Traish et al. 1977) may be accounted for by differences in assay techniques for the measurement of salt-resistant nuclear receptors. In addition, these investigators showed that estrogen binding sites were located on the nuclear matrix.

While much of this confusion regarding the physiological significance of salt-resistant receptor-estrogen complexes may be explained on the basis of methodology, recent reports from this laboratory describing a second class of nuclear estrogen binding sites has further complicated this picture. We have demonstrated that estradiol injection into immature (Eriksson et al. 1978; Clark et al. 1978) and mature ovariectomized (Markaverich, Clark 1979; Markaverich et al. 1981) rats results in the presence of two classes of nuclear binding sites for estradiol. Type I sites represent the estrogen receptor which is translocated from the cytoplasm (Eriksson et al. 1978; Clark et al. 1978) whereas type II sites appear to be a specific nuclear response to estrogen (Eriksson et al 1978; Markaverich, Clark 1979; Markaverich et al. 1981). More important are the observations that type II sites can influence the measurement of nuclear type I sites and elevations in nuclear levels of these secondary sites are dramatically influenced by the estrogen administered, duration of estrogenic exposure (acute or chronic) and the time at which these sites are measured following estrogen administration (Eriksson et al. 1978; Markaverich, Clark 1979). Furthermore, progesterone and dexamethasone antagonism of uterine growth may result from an inhibition of estrogen stimulation of nuclear type II sites rather than from interfering with the normal functions of the estrogen receptor (Markaverich et al. 1981). Consequently, if the influence of type II sites on measurements of the estrogen receptor are not carefully accounted for by saturation analysis utilizing [^{3}H]estradiol exchange (Eriksson et al. 1978;

Markaverich et al. 1981), accurate quantitation of nuclear receptor-estrogen complexes becomes very difficult. The complications mentioned above are also relevant to the use of exchange assays which are based on salt-extraction procedures (Zava et al. 1976; Garola, McGuire 1977). This commonly used technique could be subject to considerable error depending on the effects of salt-extract on type II sites. The studies presented in this paper will describe the effects of salt-extraction on nuclear type I and II estradiol binding sites and their relationship to the nuclear matrix.

EFFECTS OF SALT EXTRACTION ON NUCLEAR TYPE I AND II SITES

Earlier studies from this laboratory demonstrated that estrogen administration to immature (Eriksson et al. 1978; Clark et al. 1978) or mature ovariectomized (Markaverich, Clark 1979; Markaverich et al. 1981) rats results in two specific nuclear binding sites for estradiol. We have designated these components type I and type II. A typical saturation analysis for specific nuclear estrogen binding sites by [^{3}H]estradiol exchange is presented in Figure 1A. Although not apparent from the data in Figure 1A, the saturation plot of specifically bound [^{3}H]estradiol contains at least two estrogen binding components. These are apparent when these data are replotted by the method of Scatchard (1949) (Fig. 1B). The type I sites (Kd~40 nM) appear to be a specific nuclear response to estradiol since previously described secondary sites present in the cytosol are not depleted by an injection of estradiol (Eriksson et al. 1978; Clark et al. 1978).

The data presented in Figure 1C demonstrate the effects of 0.4 M KCl extraction on nuclear type I and type II sites 1 hr following a single injection of estradiol in the mature ovariectomized rat. The saturation curve for non-extracted nuclei in Figure 1C is the same data (specific binding) in Figure 1A and B (expressed in pmoles/uterus) demonstrating the presence of type I (1.5 pmoles/uterus) and type II (2.0 pmoles/uterus) estrogen binding sites. Notice that type I sites represent the lower portion of the saturation curve (0-8 nM) whereas type II sites bind [^{3}H]estradiol in the 16-30 nM range. At 1 hr post-injection virtually 100% of the type I sites are resistant to 0.4 M KCl extraction and very few sites are recovered in the salt-extract.

Summation of the binding data in the salt-extractable and salt-resistant fractions yields quantitative recovery of type I sites which approximates 1.5 pmoles/uterus. This value was not different from that observed in the non-extracted nuclear fractions (Fig. 1C).

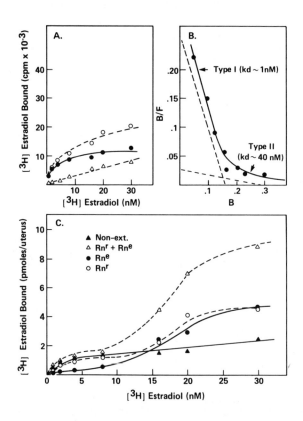

Fig. 1. (A) Saturation analysis of uterine nuclear fractions for estrogen binding sites by [^{3}H]estradiol exchange in the absence of salt. Specific binding (●) was determined by subtraction of non-specific binding (△) from the total quantity of [^{3}H]estradiol bound (o). (B) Scatchard plot of specific binding presented in Figure 1A. (C) Comparison of specific [^{3}H]estradiol binding obtained by saturation analysis of uterine nuclear fractions for estrogen binding sites following 0.4 M KCl-extraction. Abbreviations: ▲, non-extracted; △, salt-resistant plus salt-extracted sites; ●, salt-extracted sites; o, salt-resistant sites.

Conversely, salt-extraction results in a dramatic increase in the numbers of detectable type II sites as compared to those measured in the non-extracted nuclei. The salt-resistant and salt-extractable fraction contained ~4.0 pmoles/uterus type II sites each, whereas the same nuclei not exposed to 0.4 M KCl contained approximately 2 pmoles/uterus type II sites. The quantity of type II sites in the salt-resistant and salt-extractable fractions is 3- to 4-fold (~78 pmoles/uterus) greater than the amount measured in the corresponding non-extracted nuclear pellet (~2 pmoles/uterus; Fig. 1C). The results clearly demonstrate that 0.4 M KCl-extraction "opens up", or increases the numbers of type II sites 3- to 4-fold as measured by [^{3}H]estradiol exchange. In contrast, the levels of type I sites remained quantitatively stoichiometric (Fig. 1C; compare non-extracted nuclei versus salt-resistant plus extractable fractions). If, however, the influence of type II sites on

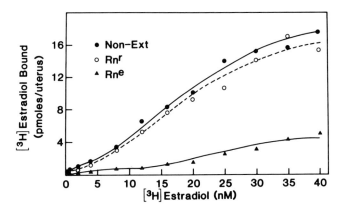

Fig. 2. Saturation analysis for specific [^{3}H]estradiol binding sites in uterine nuclear fractions obtained from estradiol (A) and estriol (B) implanted rats. Beeswax pellets (10 mg) containing 10% (w/w) estradiol or estriol were implanted (2/rat) subcutaneously and animals were sacrificed 72 hr following hormone administration. (●), non-extracted nuclei; o, salt-resistant sites; Δ, salt-extractable sites.

measurement of the estrogen receptor is not accounted for by full saturation analysis as described here (Fig. 1) the levels of type I sites, particularly at higher [^{3}H]estradiol concentrations (>10 nM) may be grossly overestimated.

To examine these relationships further, we measured nuclear levels of type I and type II sites in non-extracted, salt-resistant and salt-extractable uterine nuclear fractions following chronic exposure of the animals to estrogen administered by beeswax implant. Our rationale was that if salt-extraction was in some way mimicking the effects of estrogen on type II sites, one might not observe the "opening up" of type II sites in the salt-treated nuclear fractions from estradiol implanted rats as compared to animals receiving an estradiol injection. The data in Figure 2A demonstrate that when estradiol is administered continuously, a dramatic increase in nuclear levels of type II sites (~16 pmoles/uterus) is observed. Analysis of the specific binding in all three nuclear preparations revealed that 80% of the total nuclear type II sites (salt-resistant plus extractable ~18 pmoles/uterus) were resistant to extraction (salt-resistant ~15 pmoles/uterus) with 0.4 M KCl. Thus when estradiol was administered by implant, the type II sites appear to be locked into the nucleus in some manner which renders them resistant to high salt extraction. Identical results were obtained with an estriol implant (Fig. 2B) although the magnitude of the response to this estrogen was somewhat less than that obtained with estradiol. This is in sharp contrast to data obtained following an estradiol or estriol injection where 50% of the total quantity of type II sites are extractable with 0.4 M KCl at times longer than 1 hr post-injection (Markaverich et al. 1981).

These results describing differential effects of estrogens and salt-extraction of estimates of type I and type II sites have important implications with respect to the validity of receptor measurements in mixed binding systems. As we have previously demonstrated, elevations in the levels of cytoplasmic and nuclear type II sites can introduce errors in estrogen receptor measurements in the immature (Scatchard 1949; Rosenthal 1967) and mature ovariectomized rat uterus (Markaverich, Clark 1979; Markaverich et al. 1981) and in mouse (Watson, Clark 1980) and human (Panko et al. in press) breast tissue. As was clearly indicated by the present studies, such estimates can be further complicated by

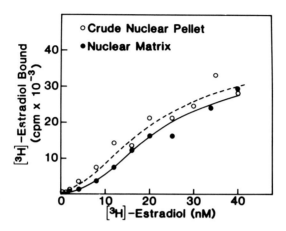

Fig. 3. Saturation analysis of specific binding of [^{3}H]estradiol to crude uterine nuclei and nuclear matrix.

exposure of tissue preparations to salt-extraction (Figs. 1 and 3). Such treatment results in a dramatic increase in the measureable levels of nuclear type II sites as compared to estimates made in low salt buffers. Conversely, quantities of type I sites following salt-extraction, remain stoichiometric (compare non-extracted nuclei vs $Rn^e + Rn^r$, Fig. 1). Thus exposure of uterine nuclei to high salt results in a 3- to 4-fold increase in detectable quantities of type II sites which inherently interferes with measurement of the estrogen receptor. Protamine sulfate precipitation and adsorption to hydroxylapatite of KCl-extracted estrogen receptors followed by [^{3}H]estradiol exchange have been employed routinely in a number of systems to reduce non-specific binding (Zava et al. 1976; Garola, McGuire 1977). Such procedures lead to erroneous results, particularly when single point exchange assays employing > 10 nM [^{3}H]estradiol are used. Under these conditions neither the effects of 0.4 M KCl on type II sites (Fig. 1) or their resulting interference with estrogen receptor (type I sites) measurements can be properly assessed.

Exposure of animals to continous estradiol by beeswax implant (Fig. 2) induces a maximum number of nuclear type II sites and additional sites are not observed following KCl-extraction (compare non-extracted vs salt-resistant and extractable nuclear fractions). This "lock in" of type II sites by continuous exposure to estrogens may represent the conversion of salt-labile estrogen binding sites to forms which are more tightly bound to nuclear components. Although the significance of these findings is not understood, the presence of these sites is correlated with true uterine growth and the possibility exists that they represent important components of the biosynthetic mechanisms which control growth.

The above data strongly suggest that full saturation analysis using a wide range of [^{3}H]estradiol concentrations is mandatory when assessing nuclear levels of receptor sites by [^{3}H]estradiol exchange (Eriksson et al. 1978; Markaverich, Clark 1979; Markaverich et al. 1981) to correct for the influence of type II sites on estimates of the estrogen receptor. In addition, quantitation of estrogen receptors by 0.4 M KCl-extraction should be used and interpreted with caution, since these techniques are subject to a number of experimental artifacts involving the estrogen adminstered and the duration of exposure. The measurement of estrogen receptors in salt-extracts of target cell nuclei was proposed by Zava et al. (1976) as a method which lowered non-specific binding in exchange assays (Garola, McGuire 1977). Although this method does decrease the non-specific binding, it is subject to the errors described above and is likely to lead to incorrect conclusions if greater than 2-4 nM [^{3}H]estradiol is used in exchange assays. The original exchange method for the measurement of nuclear bound estrogen receptors, coupled with the modifications which are presented here and elsewhere (Markaverich et al. 1981), enable an investigator to measure both salt-extractable and resistant sites. Thus no error will be introduced by variable quantities of salt-insoluble receptor sites. In addition, the qualitative and quantitative relationship between type I and type II sites can be assessed.

NUCLEAR TYPE II SITES AND NUCLEAR MATRIX

The resistance of the nuclear type II site to 0.4 M KCl extraction (Fig. 2) suggests this site is tightly associated with nuclear structures. Previous reports (Barrack, Coffey 1980) have demonstrated that receptors for androgens and estrogens in uterus, liver and ventral prostate are present on nuclear matrix. These investigators have also shown that the matrix represents an intranuclear protein network which is essentially devoid (98-99%) of RNA, DNA, lipid and chromosomal protein and may serve as an attachment site for DNA replication (Pardoll et al. 1980).

The data presented in Figure 3 suggest type II sites in the rat uterus are also associated with the nuclear matrix of this tissue. In these experiments, nuclear matrix was prepared from uterine nuclei obtained from estradiol-implanted rats (2 mg for 96 hr) essentially as described by Barrack et al. (1977). Briefly, uteri were homogenized in high magnesium buffer (10 mM Tris; 5 mM $MgCl_2$, pH 7.4) and the homogenate centrifuged at 800 x g to obtain the crude nuclear pellet. The pellet was washed three times by resuspension and centrifugation (800 x g for 20 min) in high magnesium buffer and split into two equal aliquots. One aliquot was assayed directly for nuclear type II sites by [^{3}H]estradiol exchange (Fig. 3; o) under conditions which are optimum (4° for 60 min) for the measurement of this site. The remaining nuclei (Fig. 3; o) were extracted three times with 2 M NaCl, washed with 1% Triton X-100 and also assayed for nuclear type II sites by [^{3}H]estradiol exchange. Analysis of these data revealed that 95-98% of the nuclear type II sites were resistant to these extraction procedures and are presumably associated with nuclear matrix, even though this matrix preparation was devoid of 90-95% of the total nuclear DNA. Preliminary data also suggest these sites are resistant to DNase digestion and are dramatically reduced by trypsin (data not shown), further supporting their association with nuclear matrix.

As stated earlier, the precise role of nuclear type II sites in estrogen action is unclear and many questions regarding their function remain to be answered. Close association of these sites with nuclear matrix and a possible role for the regulation of DNA replication seems tempting but at the present is totally speculative.

Anderson JN, Clark JH, Peck EJ, Jr (1972). The relationship between nuclear receptor estrogen binding and uterotrophic responses. Biochem Biophys Res Comm 48:1460.

Anderson, JN, Peck, EJ, Jr, Clark, JH (1973). Nuclear receptor estrogen complex: relationship between concentration and early uterotrophic responses. Endocrinology 92:1488.

Barrack ER, Hawkins EF, Allen SL, Hicks LL and Coffey DS (1977). Concepts related to salt-resistant estradiol receptors in rat uterine nuclei, nuclear matrix. Biochem Biophys Res Commun 79:829.

Barrack ER and Coffey DS (1980). The specific binding of estrogens and androgens to the nuclear matrix of sex hormone responsive tissues. J Biol Chem 255:7265.

Clark JH and Peck EJ, Jr (1976). Nuclear retention of receptor-estrogen complex and nuclear acceptor sites. Nature 260:635.

Clark JH, Hardin JW, Upchurch S, Eriksson H. (1978). Heterogeneity of estrogen binding sites in the cytosol of the rat uterus. J Biol Chem 253:2630.

Clark JH, Peck EJ, Jr (1979). "Female sex steroids: receptors and function." New York: Springer-Verlag.

Erikkson H, Upchurch S, Hardin JW, Peck EJ, Jr, Clark JH (1978) Heterogeneity of estrogen receptors in the cytosol and nuclear fraction of the rat uterus. Biochem Biophys Res Commun 81:1.

Garola RE, McGuire WL (1977). An improved assay for nuclear estrogen receptor in experimental and human breast cancer. Cancer Res. 37:3333.

Juliano JV, Stancel GM (1976). Estrogen receptors in the rat uterus. Retention of hormone-receptor complex. Biochemistry 15:916.

Markaverich BM, Clark JH (1979). Two binding sites for estradiol in rat uterine nuclei: relationships to uterotrophic response. Endocrinology 105:1458.

Markaverich BM, Upchurch S, Clark JH (1981). Progesterone and dexamethasone antagonism of uterine growth: a role for a second nuclear binding site for estradiol in estrogen action. J Steroid Biochem 14:125.

Markaverich BM, Upchurch S, Clark JH (1981). Effects of salt extraction on the quantitation of nuclear estrogen receptors: interference by secondary estrogen binding sites. J Receptor Res. 1:415.

Panko, WB, Watson CS, Clark JH The presence of a second, specific estrogen binding site in human breast cancer. Cancer Res., in press.

Pardoll, DM, Vogelstein B, Coffey DS (1980). A fixed site DNA replication in eucaryotic cells. Cell 19:527.

Rosenthal HE (1967). A graphic method for the determination and presentation of binding parameters in a complex system. Anal Biochem 20:525.

Ruh TS, Baudendistel LJ (1977). Different nuclear binding sites for anti-estrogen and estrogen receptor complexes. Endocrinology 100:420.

Scatchard G (1949). The attractions of proteins for small molecules and ions. Ann NY Acad Sci 51:660.

Traish AM, Muller RE, Wotiz HH (1977). Binding of estrogen receptor to uterine nuclei. Salt-extractable versus salt-resistant receptor estrogen complexes. J Biol Chem 252:6823.

Watson, CS, Clark JH (1980). Heterogeneity of estrogen binding sites in mouse mammary cancer. J Receptor Research 1:91.

Zava DT, Harrington NY, McGuire WL (1976). Nuclear estradiol receptor in the adult rat uterus: a new exchange assay. Biochemistry 15: 4292.

A MODEL FOR NUCLEOCYTOPLASMIC TRANSPORT OF RIBONUCLEOPROTEIN PARTICLES

Gary A. Clawson, Ph.D. and Edward A. Smuckler, M.D., Ph.D.

Department of Pathology
University of California School of Medicine
San Francisco, CA 94143

SUMMARY

Some aspects of the transport of ribonucleoprotein (RNP) particles from nucleus to cytoplasm have been derived from in vitro assays employing isolated nuclei. The transport process has an apparent activation energy of 13 kcal/mol, shows an Arrhenius relationship without evidence of a transition between 35 and 0°C, and requires hydrolysis of one high-energy phosphate bond per nucleotide in transported RNA. All of these analyses of the process are independent of the type of RNP particle [ribosomal (rRNP) or messenger (mRNP)] transported. A serious conceptual difficulty arises when the size of the transported particles is considered. They must presumably travel through an aqueous channel in the nuclear envelope (since lipid phase transitions do not appear in the Arrhenius graphs), but all of the transported particles are too large to pass through the nuclear pore complexes. In reviewing what is known about RNP structure, we find common features which suggest the following model:

1) The RNA chain is exposed at the surface of the particles.
2) Small, local regions of particle structure "unfold."
3) These unfolded (linear) segments of RNA interact with a translocation mechanism containing a nucleoside triphosphatase.
4) The RNA chain is linearly translocated through the pore channel the length of one nucleotide for each high-energy phosphate bond hydrolyzed.

5) The particle then refolds outside of the nuclear envelope.

The mechanism by which RNA is transferred from the nuclear interior to the cytoplasm is a process integral to cellular function. Due to the inherent complexities of this transport in vivo, in vitro assays employing isolated nuclei have been used to define the mechanisms involved; the process involves maturation of large nuclear RNA transcripts to much smaller molecules, and movement of cellular RNA complexed with protein in RNP particles (Roy et al. 1979; Chistick et al. 1979; Sato et al. 1977).

Ribosomal, messenger, and transfer RNA (rRNA, mRNA, tRNA) maturation occur with considerable fidelity with isolated nuclei in vitro. tRNA is correctly processed in vitro (O'Farrell et al. 1978). When ribosomal proteins are present in the incubation brei, these proteins are rapidly taken up by isolated rat-liver nuclei (Bolla et al. 1977); they complex with 45S pre-rRNA RNA to form 80S pre-RNP particles in the nucleolus, analogous to formation of 80S-pre-RNP in the nucleolus in vivo (Matsuura et al. 1974). The RNA in these particles is then processed to mature 28S and 18S rRNA species, which are found in the nucleoplasm as components of 60S and 40S ribosomal subunits (Bolla et al. 1977). When adenosine triphosphate (ATP) is included with cell sap in the incubation mixture, the 60S and 40S ribosomal subunits are the predominant form of RNP transported from the nucleus to surrogate cytoplasm (Sato et al. 1977; Racevskis, Webb 1974; Schumm, Webb 1974, 1981). The 60S and 40S ribosomal subunits released in vitro have buoyant densities (respectively) of 1.61 and 1.56 g/cm^3 (Sato et al. 1977), characteristic of cytoplasmic ribosomal subunits.

Maturation of heterogeneous nuclear RNA (hnRNA) to mRNA has recently been demonstrated with specific probes. Goldenberg and Raskas (1980) and Yang et al. (1981) have demonstrated that correct "splicing" of adenovirus-2 hnRNA (to mRNA) occurs with isolated nuclei, a process which involves removal of an intervening sequence and subsequent ligation of two disjoint sequences.

When isolated nuclei are incubated in mixtures containing ATP (but no cell sap) predominantly messenger RNP are released from them. They display many similarities with mRNP and pre-mRNP found in vivo (Deimel et al. 1977; Howard

1978; Jain et al. 1979). The mRNP released in vitro sediment through sucrose gradients at about 40-45S (Ishikawa et al. 1969; Raskas 1971; Smuckler, Koplitz 1974), have buoyant density (Ishikawa et al. 1969; Raskas 1971) characteristic of cytoplasmic mRNP (1.40 g/cm^3), and can be incorporated into polysomes in reconstructed systems (Ishikawa et al. 1970a). They contain RNA similar in size and base composition (Ishikawa et al. 1970b) to cytoplasmic RNA, and which can direct protein synthesis in vitro (Ishikawa et al. 1970b). Poly(adenylic acid) [poly(A)]-containing RNA released in vitro hybridizes (under stringent conditions) to complementary DNA (cDNA) probes transcribed from cytoplasmic poly(A)-containing RNA to the same extent as the homologous cytoplasmic population (Clawson, Smuckler in press), indicating a considerable sequence homology between these populations.

Some aspects of the energetics of the transport process can be inferred from in vitro assays: i) RNP transport displays an apparent activation energy of 13 kcal/mol (Clawson, Smuckler 1978; see also Patterson et al. 1981), a value which is equivalent to that for adenosine triphosphatase (ATPase) activity (Dean, Tanford 1978; Clawson et al. 1980). ii) Arrhenius analyses of initial linear rates of transport yield lines of continuous slopes over the temperature domain of 0-35°C, and detergent treatment does not affect RNA transport (Clawson, Smuckler 1978; Patterson et al. 1981); thus, the translocation mechanism is apparently exposed to an aqueous environment and not influenced by lipid. Since the function of translocation is to transport RNP (RNA) through a membranous envelope, we must assume an aqueous channel exists in the nuclear envelope, since all other exits would traverse lipid barriers. Whether or not this channel corresponds to the well-described nuclear pore complex is unknown. iii) Approximately one high-energy phosphate bond is hydrolyzed in the facilitated transport of each nucleotide in RNA transported in vitro (Clawson et al. 1978). These high-energy bonds are hydrolyzed by a nucleoside triphosphatase activity associated with the nuclear envelope (Agutter et al. 1979; Vorbrodt, Maul 1980; Clawson et al. 1980). iv) The activation is unchanged by the presence or absence of cell sap (Clawson, Smuckler 1978); however, predominantly 60S and 40S ribosomal subunits are transported in the presence of cell sap, whereas only mRNP are transported without cell sap (Sato et al. 1977). Since the RNA is transported with the same energy requirement per nucleotide in both types of mixture, the RNA component, not the protein component or three-dimensional particle structure, must dictate the energy requirement. The great differences in the RNA-protein ratios in the different types of RNP (the buoyant densities of the mRNP and the rRNP differ greatly) and the dissimilarities among the particles in three-dimensional structure necessitate this conclusion.

Since the translocation mechanism is exposed to the aqueous phase and interacts with transportable RNA on a per nucleotide basis, it seems reasonable to suggest that the RNA in RNP particles must be exposed on the surface of the particles (see below). However, definition of RNP particles in the nucleus, in surrogate cytoplasm, and in vivo, leads to significant conceptual difficulties concerning their transport. Although electron microscopic studies show nuclear pore complexes with an inner diameter of about 66 nm, a much smaller channel is generally observed (Gall 1967), and pore complexes have a much smaller functional pore radius (Feldherr, Pomerantz 1978). The most refined estimate of patent pore diameter is 9 nm, based on the isotopic tracer diffusion studies of Paine et al. (1975). But Nonomura et al. (1971) have shown that ribosomal subunits (both large and small) are far too large to pass through 9-nm pore channels. Similarly, Lukanidin et al. (1973) have shown that nuclear mRNP are about 20 nm in diameter, and Dubochet et al. (1973) have demonstrated that cytoplasmic mRNP are also much greater than 9 nm in diameter. Analogous considerations hold for mRNP transported in vitro, which are dimensionally similar to the small ribosomal subunits. Further, measurements such as those of Pilz et al. (1970) clearly indicate that even mature tRNA would require from a few hours a day to reach nucleocytoplasmic equilibrium (based on the data of Paine et al. 1975), although this is clearly not the case in vivo.

A MODEL

The mechanisms with which we propose to explain these diverse data involves the following ideas: 1) the RNA chain is exposed at the surface of the RNP; 2) small, local regions of RNP structure "unfold"; 3) these unfolded (linear) segments of RNA interact with a translocation mechanism containing a nucleoside triphosphatase; 4) the RNA chain is linearly translocated through an aqueous channel the length of one nucleotide for each high-energy phosphate bond hydrolyzed; and 5) the RNP particle then refolds outside of the nuclear envelope. This sequence is consistent with the results of our bookkeeping in the in vitro system we have employed, and with the structural features of RNP, some of which are discussed below.

In the model of the ribosome structure proposed by Cotter et al. (1967), the ribosomal surface is largely composed of RNA, with helical regions protruding outward and with spherical protein domains (about 3 nm in diameter) attached to nonhelical RNA segments inside the subunits. Cox (1969), along with Bonanou (Cox, Bonanou 1969), later presented a more detailed "horsehoe and cap" model, which retained protein domains 3-4 nm in diameter, and which had hairpin loops (two loops per protein domain, each composed of about seven base pairs and nine unpaired residues) covering the surface of each ribosomal subunit. Studies of endonuclease digestion by Spencer and Walker (1971) support the hairpin-loop hypothesis, and further studies by Lind et al. (1975) suggest that the single-stranded hairpin loops consist of pyrimidine-rich clusters. Thus, the main features of the model of the ribosome, as it has evolved, are that the RNA is coiled into hairpin loops and is located at the ribosomal surface, and that the protein subunits (which are tightly linked to single-stranded segments of rRNA) are 4 nm or less in diameter. The existence of small domains of similar secondary structures would have tremendous energetic advantages. The linearization of small hairpin segments would require much less energy than the unfolding or linearization of large portions of the rRNA molecules. In addition, the single-stranded hairpin-loop segments are composed of pyrimidines, which convey a great deal of flexibility to such structures.

The demonstration by Matsuura et al. (1974) that nucleolar RNP particles (sedimenting at 80-100S) containing rRNA and precursors of rRNA also exist as pleomorphic, rod-like, and filamentous structures with nodular thickenings, provides experimental support for the postulated unfolding.

There are basic similarities between the structure of the ribosomal subunits and that of informofers. Hairpin-like structures have been found in pre-mRNA in nuclear RNP particles (Molnar et al. 1975); apparently these structures are formed directly after transcription and participate in mRNA processing. Lukanidin et al. (1973a) proposed that the

RNA chains in nuclear RNP particles are wrapped around the outside of the particles and thus are available for interaction with a translocation mechanism. Lukanidin et al. (1973b) found a 30S nuclear RNP composed of RNA associated with many copies of a homogenous protein termed "informatin." The particles were self-assembling in reconstitution experiments, providing direct evidence that RNP packaging is not an energy-requiring step. Billings and Martin (1978) have reported similar results.

The hairpin loops in Cox's model of the ribosome have the general features of those described by other workers (Pilz et al. 1970; Cramer 1971) for tRNA structure. Therefore, as required by our model, in all transported material, the RNA chain is exposed to the aqueous phase and has single-stranded hairpin loops, whether it is free in solution (as in tRNA precursor) or localized on the surfaces of nuclear RNP particles (as in mRNP and ribosomal units). The translocation mechanism presumably recognizes specific conformational aspects of RNA or mature RNP particles, unfolding local regions of RNA (RNP) structure and translocating the RNA chain linearly with associated protein domains. On the basis of our model, the specificity of transport must lie in the processes occurring before interaction with the translocation mechanism, since it is doubtful that the selective mechanism could operate with suitable fidelity through interaction with short segments of many diverse RNA chains. An interesting possibility is that small RNA species (such as 5S ribosomal RNA and the small species shown to be present in mRNP structures (Deimel et al. 1977; Howard 1978) fulfill such a function, the "reusable" small RNA species being shuttled back and forth. This idea is consistent with the observed behavior of some RNA species (Wise, Goldstein 1973), and with the reentry of small RNA observed in rat-liver nuclei (Clawson, Smuckler 1978).

ACKNOWLEDGMENT

Supported by grant CA 21141 from the NIH.

Agutter P, McCaldin B, McArdle H (1979). Biochem J 182:811.
Billings P, Martin T (1978). J Cell Biol 79:375a (absrt).
Bolla R, Roth H, Weissbach H, Brot N (1977). J Biol Chem 252:721.

Chistick M, Brennessel B, Biswas D (1979). Biochem Biophys Res Commun 91:1109.
Clawson G, James J, Woo C, Friend D, Moody D, Smuckler E (1980). Biochemistry 19:2748.
Clawson G, Koplitz M, Castler-Schechter B, Smuckler E (1978). Biochemistry 17:3747.
Clawson G, Smuckler E (1978). Proc Natl Acad Sci USA 75:5400.
Cotter R, McPhie P, Gratzer W (1967). Nature (London) 216:864.
Cox R (1969). Biochem J 114:753.
Cox R, Bonanou S (1969). Biochem J 114:769.
Cramer F (1971). Prog Nuc Acid Res Molec Biol 11:391.
Dean W, Tanford C (1978). Biochemistry 17:1683.
Deimel B, Louis C, Sekeris C (1977). FEBS Lett 73:80.
Dubochet J, Morel C, LeBleu B, Merzberg M (1973). Eur J Biochem 36:465.
Feldherr C, Pomerantz J (1978). J Cell Biol 78:168.
Gall J (1967). J Cell Biol 32:391.
Goldenberg C, Raskas H (1980). Biochemistry 19:2719.
Howard F (1978). Biochemistry 17:2719.
Ishikawa K, Kuroda C, Ogata K (1969). Biochim Biophys Acta 179:316.
Ishikawa K, Ueki M, Nagai K, Ogata K (1970a). Biochim Biophys Acta 213:542.
Ishikawa K, Kuroda C, Ueki M, Ogata K (1970b). Biochim Biophys Acta 213:495.
Jain S, Pluskal M, Sarkar S (1979). FEBS Lett 97:84.
Lind A, Villems R, Saarma M (1975). Eur J Biochem 51:529.
Lukanidin E, Georgiev G, Williamson R (1973b). Molekulyarnaya Biologiya 7:264.
Lukanidin E, Kul'Guskii V, Aitkhozhina N, Komaromi L, Tikhonenko A, Georgiev G (1973a). Molekulyarnaya Biologiya 7:360.
Matsuura S, Morimoto T, Tashiro Y, Higashinakagawa T, Muramatsu M (1974). J Cell Biol 63:629.
Molnar J, Besson J, Samarina O (1975). Mol Biol Rpts 2:11.
Nonomura Y, Blobel G, Sabatini D (1971). J Mol Biol 60:303.
O'Farrell P, Cordell B, Valenzuela P, Rutter W, Goodman H (1978). Nature 274:438.
Paine P, Moore L, Horowitz S (1975). Nature (London) 254:109.
Patterson R, Lyerly M, Stuart S (1981). Cell Biol Int Rpts 5:27.
Pilz I, Kratky O, Cramer F, Harr F, von der Schlimme E (1970). Eur J Biochem 15:401.

Racevskis J and Webb T (1974). Eur J Biochem 49:93.
Raskas J (1971). Nature New Biol 233:134.
Roy R, Lau A, Munro H, Baliga B, Sarkar S (1979). Proc Natl Acad Sci USA 76:1751.
Sato T, Ishikawa K, Ogata K (1977). Biochim Biophys Acta 474:536.
Schumm D, Webb T (1974). Biochem J 139:191.
Schumm D, Webb T (1978). J Biol Chem 253:8513.
Smuckler E, Koplitz M (1974). Cancer Res 34:827.
Spencer M, Walker I (1971). Eur J Biochem 19:451.
Vorbrodt A, Maul G (1980). J Histochem Cytochem 28:27.
Wise G, Goldstein L (1973). J Cell Biol 56:129.
Yang V, Lerner M, Steitz J, Flint S (1981). Proc Natl Acad Sci USA 78:1371.

THE INTERACTION OF NUCLEAR REACTANT DRUGS WITH THE NUCLEAR MEMBRANE AND NUCLEAR MATRIX

Kenneth D. Tew

Division of Medical Oncology and Department of Biochemistry
Lombardi Cancer Center
Georgetown University
Washington, D.C. 20007

A precise molecular rationale for the induction of cell death by nuclear reactant drugs remains to be established. The nuclear membrane and matrix are rich sources of nucleophilic sites and as such are shown to be preferential targets for chloroethylnitrosoureas, a major class of antitumor agents. Because of the imputed importance of both the membrane and matrix to nuclear structure and function, the potential for drugs to interfere with or modify these structures is reviewed. There is evidence that the nuclear matrix will play an integral role in determining the fate of a cell which has been subjected to a cytotoxic drug.

NUCLEAR TARGET SPECIFICITY

A broad spectrum of anticancer agents may be classified as "nuclear-reactant" drugs. Such compounds are capable of producing chemical moieties which can interact covalently with cellular nucleophilic species. The nuclear membrane and matrix, as well as associated chromatin components, are rich sources of such macromolecular nucleophiles and therefore provide ample targets for drugs such as nitrosoureas, nitrogen mustards, anthracycline antibiotics, vinca alkaloids and steroid hormones. These drugs play a major role in the control and treatment of human neoplastic diseases and yet there is no definitive mechanism which links their pharmacological properties with drug-induced cell death. It remains paradigmatic that nuclear reactant drugs initiate cytotoxicity through a modification of DNA. Exactly how and when a cell dies is subject to evaluation. Although it is generally accepted that proteins, phospholipids, cytoplasmic and nuclear RNA are sufficiently redundant and possess high enough turnover rates to be relatively unaffected by these drugs, this may not be true for all cells and for all drugs. In addition, it is probable that not all DNA is of equal significance as a potential target, since there can be large amounts of inactive and redundant material within cells. Some or all of the factors outlined below are pertinent to considerations of the nuclear membrane or matrix as important drug targets: (i) Both present structural barriers through which drug species must pass to reach chromatin. (ii) They contain large numbers of nucleophilic macromolecular species. (iii) Their structure is dynamic rather than fixed. (iv) The ultimate target, chromatin, is contiguous with both structures. (v) There is flexibility in the ratio of phospholipids and proteins. (vi) There is a possibility for morphological variability to exist between different cell types. (vii) Cell death can be correlated with a loss of nuclear membrane negative charge. Evidence will be presented that both the membrane and the matrix are preferential targets for chloroethylnitrosoureas, with disproportionate alkylation and carbamoylation of structural macromolecules. Drug interactions with the nuclear membrane and matrix may prove to be important mediators of drug-induced cytotoxicity.

THE CONCEPT OF CELL DEATH

It is as difficult to define the precise moment at which a cell dies as to determine the exact drug-induced lesion which is responsible for its demise. Figures 1 and 2 illustrate a transition of a normal HeLa cell into nuclear structural degeneration and cell death. In Figure 1 there is a normal distribution of hetero- and euchromatin within the bilayered nuclear membrane. Nucleolar integrity is apparent, as also is the contiguous nature of the chromatin with the inner nuclear membrane (INM). In the dying cell (Fig. 2) evidence of random chromatin clumping and nucleolar disintegration is apparent. Additionally, the attachments of chromatin to the INM are broken or in the process of disassociating. The bilayer integrity of the membrane has been affected with the INM separating from the outer nuclear membrane (ONM). The precise reason for this separation is unclear, but must involve some drastic modifications of the localized charge properties of the membrane and associated macromolecules. Whether the breakdown of INM-chromatin association is the cause or consequence of cell death is unclear. The possibility that nuclear-reactant drugs cause cell death by interfering with charge relationships at the level of the nuclear membrane (and by connotation, the matrix) is worthy of further investigation.

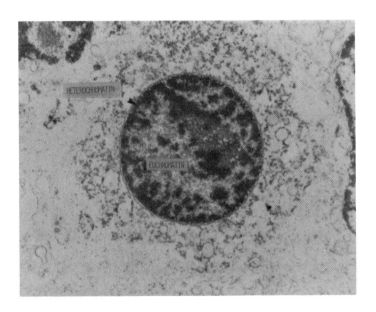

Fig. 1. Untreated HeLa cell (15,000X). Electron-translucent euchromatin is surrounded by the electron-dense, INM-associated heterochromatin. Semithin section of Epon-embedded, osmicated pellets were dehydrated with acetone and stained with aqueous uranyl acetate.

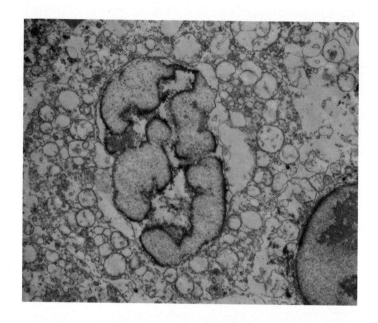

Fig. 2. A dying HeLa cell pretreated with 1 µM hydrocortisone (HC) (22 hr) and 0.6 mM CCNU (2 hr) (15,000X). The breakdown of nuclear structural integrity is apparent. The INM has separated from the ONM in some places. The chromatin has formed irregular clumps which differ dramatically from the normal chromatin organization (Fig. 1). Nucleolar disintegration is also apparent (from Tew et al. 1980).

PHARMACOLOGY OF NUCLEAR REACTANT DRUGS

Many books and reviews cover the basic pharmacology of alkylating agents and related drugs (Price 1975; Pratt, Ruddon 1979). Similarly, nitrosoureas have been covered in detail elsewhere (Prestayko et al. 1981; Tew et al. 1981). Briefly, chloroethylnitrosoureas decompose to yield alkyldiazohydroxide and isocyanate moieties which are capable of alkylating and carbamoylating nucleophilic sites, examples of which are shown in Figure 3. This class of cytotoxic agent can also interact with amino, carboxyl, sulfhydryl or imidazole moieties in other proteins and nucleic acids which can participate in the formation of any of the following drug lesions: (i) monofunctional adducts, (ii) crosslinkage of nucleic acids (inter- or intrastrand), (iii) crosslinkage of nucleic acids with proteins, (iv) strand breakage, (v) carbamoylation of proteins.

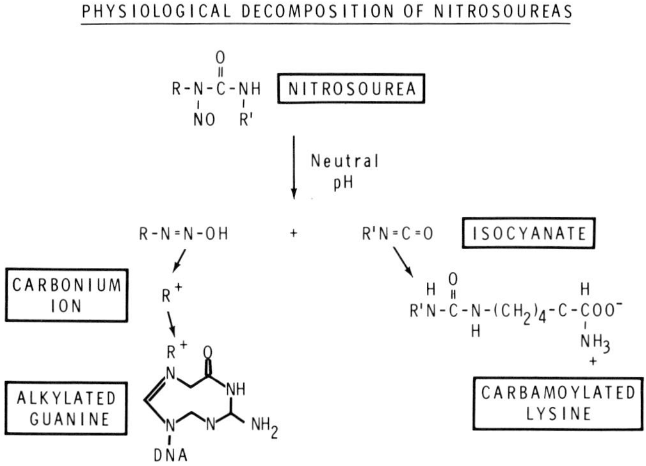

PHYSIOLOGICAL DECOMPOSITION OF NITROSOUREAS

$$R-N-\underset{\substack{\| \\ NO}}{\overset{O}{C}}-NH \quad \boxed{NITROSOUREA}$$

↓ Neutral pH

$R-N=N-OH \quad + \quad R'N=C=O \quad \boxed{ISOCYANATE}$

$\boxed{\text{CARBONIUM ION}} \quad R^+$

$R'N-\underset{H}{\overset{O}{\underset{\|}{C}}}-N-(CH_2)_4-\underset{NH_3}{\overset{H}{\underset{|}{C}}}-COO^-$

R^+

$\boxed{\text{ALKYLATED GUANINE}}$ (structure with DNA)

$\boxed{\text{CARBAMOYLATED LYSINE}}$

Fig. 3. Schematic representation of nitrosourea decomposition.

NUCLEAR MEMBRANE AND MATRIX AS SPECIFIC DRUG TARGETS

Studies using cultured HeLa S3 cells have shown that both the membrane and the matrix are preferential targets for two chloroethylnitrosoureas, 1-(2-chloroethyl)-3-cyclohexyl-1-nitrosourea (CCNU) and 2-[3-(2-chloroethyl)-3-nitrosoureido]-2-deoxy-D-glucopyranose (chlorozotocin:CLZ) (Tew et al. 1980; Wang et al. 1981; Tew et al. 1981). Using radiolabeled nitrosoureas it was possible to show that approximately one third of the total nuclear-associated drug was bound to the membrane and matrix (Table 1).

Considering the small percentage of nuclear macromolecules which constitute these structures, such drug interac-

Table 1

Comparative Nitrosourea Interactions with the Nuclear Matrix

Drug	Nuclear membrane	Nuclear matrix
CLZ alkylation	38.7	26.7
CCNU alkylation	33.4	31.3
CCNU carbamoylation	55.0	33.1

Data are expressed as percentage of total nuclear bound drug.

tions represent high levels when compared to the bulk chromatin. Further attempts to fractionate the matrix have suggested that the ribonuclearprotein and fibrillar elements of the matrix are alkylated and carbamoylated equally effectively (Tew et al. 1981). These data indicate that a large number of nucleophilic species exist in both the membrane and the matrix. The integral location of these nucleophilic sites will have served to expose them to drug species transported into the nucleus, thereby assigning them as preferential targets. Other studies using rat liver and lung have demonstrated a preferential binding of benzopyrene to non-chromatin matrix fractions (Hemminki, Vainio 1979).

CORTICOSTEROID-INDUCED ALTERATIONS OF THE MATRIX

It is possible to induce a variety of nuclear changes in target cells by exposing them to steroids. The localization of certain estrogen and androgen receptors on the matrix of sex-hormone responsive tissues (Barrack, Coffey 1980) is indicative of the fact that the hormone response may be mediated through (or is concomitant with) matrix rearrangement. Since corticosteroids and alkylating agents are used in combination modality for the treatment of a number of neoplastic disease (Talley 1973), we have examined in HeLa S3 cells the molecular consequences of nitrosoureas in combination with HC. HC receptors were estimated at 8 fmol/µg DNA (Tew et al. 1981b) and 22 hr pretreatment at 1 µM resulted in an increase in nuclear size and a dispersion of peripheral heterochromatin (compare Fig. 4 to Fig. 1).

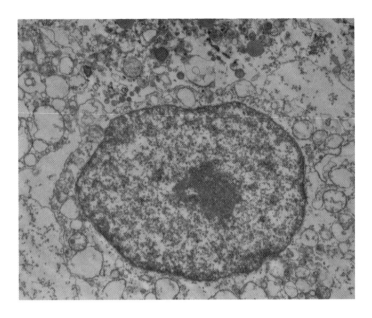

Fig. 4. HeLa cell pretreated for 22 hr with 1 µM HC (15,000X). Membrane-associated HC has become more diffuse and less electron-dense. The nucleus has swelled to approximately twice its pretreatment size (from Tew et al. 1980b).

Concomitant with these morphological effects was an approximately 10-fold stimulation of transcriptional activity as measured by uridine incorporation (Tew et al. 1980b). The extended nature of the chromatin was found to confer a different sensitivity to micrococcal nuclease digestion as measured by sucrose gradient centrifugation (Fig. 5). The production of nucleosome peaks was confirmed by chromatin gel electrophoresis. The hormone dependent shoulder may be the result of ribonuclear material, stimulated by HC, sedimenting in this range. In addition to the induction of chromatin disaggregation, an increased alkylation (CCNU and CLZ) and carbamoylation (CCNU) of transcriptional chromatin was found to occur (Tew et al. 1980b). By measuring in vitro colony-forming ability, combinations of HC with CLZ or CCNU were found to induce cytotoxicity synergistically, when

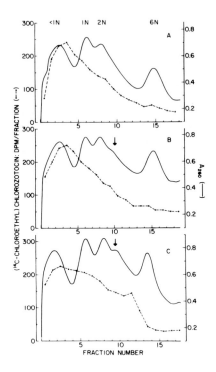

Fig. 5. 10-30% sucrose density gradient separation of micrococcal nuclease-digested chromatin from HeLa cells with and without HC pretreatment. All samples received 30 μm ([^{14}C]chlorethyl) CLZ for 2 hr prior to digestion. The same units of A_{260} were added to each gradient. Approximately 20% of the bulk chromatin was digested by the nuclease.

Panel A: No hormone treatment.

Panel B: 1.0 μm HC 22 hr prior to digestion.

Panel C: 2.0 μm HC 22 hr prior to digestion.

<1N is submonomer material; 1N, monomer peak; 2N, dimer peak; 6N, hexamer peak. The arrows indicate a hormone concentration-dependent shoulder. (---) Represents TCA-precipitable dpm of ([^{14}C]chloroethyl)CLZ/fraction (from Tew et al. 1980).

compared to the nitrosourea alone (Tew et al. 1981b). The increased cytotoxic potential of this combination is presumably a function of the HC-induced alteration of drug binding within the nucleus. The disaggregation of chromatin and nuclear swelling may have increased the number of nuclear matrix and associated chromatin sites which are available for drug interactions. Modulation of the cytotoxic properties of chlorambucil by prednisolone have been reported (Wilkinson et al. 1979) with similar nuclear swelling and chromatin rearrangement. The precise role of the matrix as a cytotoxic target remains equivocal. However, its involvement in the dynamic aspects of hormone-induced nuclear swelling predicates structural modifications which would modify drug binding and may reduce the cell's ability to repair damage and recover from the drug insult.

DRUG RESISTANCE AND THE NUCLEAR MATRIX

One of the major obstacles in the successful treatment of human malignancies is the ultimate selection of a tumor cell population which is refractory to conventional drugs. Most solid tumors are composed of heterogeneous cell populations and there is evidence that such heterogeneity contributes, in part, to the variable response of tumors to chemotherapy (Barranco et al. 1973; Heppner et al. 1978). The presence of specific clones of cells has been demonstrated karyotypically (Mitelman et al. 1972), as has the development of drug resistance, which can be linked with karyotypic heterogeneity (Houghton et al. 1981) and chromosome marker regions, such as homogeneously stained regions (HSR) (Biedler, Spengler 1976; Kovacs 1979). Studies in vitro using a rat carcinoma cell line (Walker 256) with acquired resistance to bifunctional nitrogen mustards has confirmed that both total number and marker differences exist between the chromosomes of the resistant and sensitive cell lines. Table 2 summarizes some of the nuclear structural differences between these cell lines.

Since most of the nuclear envelope and pore complex is salvaged and recycled within the chromosomes at mitosis (Maul 1977; Conner et al. 1980), it was not surprising that preliminary high-pressure liquid chromatography separation of matrix proteins indicated the presence of an extra polypeptide within the nuclear matrix of WX cells (Wang, Tew 1981). Whether this was a novel polypeptide or a modifica-

Table 2

Walker Cell Chromosome Analysis

	Walker-resistant	Walker-sensitive
Polyploid percentage	10	10
Modal chromosome number	56	62
DNA ratio	1	1.25
RNA ratio	1	1.4
No. of marker chromosomes	9-11	8-11
Marker Regions		
Large metacentric	++	++
Large acrocentric	-	+
Acrocentric H2 with terminal arm HSR	+ short HSR	+ long HSR
Submetacentric H1	-	+
Submetacentric H3	+	-
Submetacentric variable banding	++	++

tion of an existing protein has not been determined. It would be of interest to speculate that the acquisition of resistance depended upon the alterations of nuclear structure controlled by the matrix. Whether such matrix differences are prerequisite for the expression of drug resistance or just a tangential result of drug resistance may be relevant to the rational design of membrane or matrix target-specific drugs. In addition, it may be interesting to study the nuclear matrix composition of subpopulations of heterogeneous tumor cells, such as human malignant gliomas. These tumors possess a high degree of nuclear instability and karyotypic heterogeneity (Shapiro et al. 1981) which may be expressed through distinguishable membrane and matrix differences.

DRUG EFFECTS ON NUCLEAR MEMBRANE CHARGE

The overall electrophoretic charge (or zeta potential) of a membrane is a composite of the charges of individual macromolecules contained within and close to the surface of the membrane. The treatment of cell cultures with X-irradiation produces nuclear invaginations and dose-dependent alterations of the electrophoretic mobility of isolated nuclei (Szekely et al. 1980). A loss of the negative charge of the

Table 3

	Walker-resistant	Walker-sensitive
Whole Cell		
Control	0.97 ± 0.12	1.58 ± 0.23*
100 μM CCNU	0.95 ± 0.11	1.04 ± 0.13
Nucleus		
Control	1.23 ± 0.17†	1.32 ± 0.24*
100 μM CCNU	0.96 ± 0.23	0.87 ± 0.14

* $p = <0.001$ † $p = <0.02$

The electrophoretic mobilities of the plasma and nuclear membranes of Walker-resistant (WR) and -sensitive (WS) cells expressed in μm/sec/cm/volt. All values are negative and represent the mean ± SD for at least 30 experiments. Measurements were made on a Zeta-meter (Zeter Meter Inc., NY) using whole cells or triton-isolated nuclei suspended in 0.3 M sucrose, 40 mM sodium phosphate pH 7.2. Treatment with CCNU was for 2 hr, after which unreacted drug was removed by washing and 4 hr further incubation at 37°C was permitted. 100 μM CCNU is a ID_{100} concentration for both WR and WS. Statistical differences between untreated and drug-treated groups were judged on the basis of student $\underline{t}$ tests.

nuclear membrane has been correlated with decreased cell survival, as judged by colony-forming ability (Sato et al. 1975). A definitive explanation for the formation of these invaginations has not been forthcoming. X-rays are known to produce strand breakage in nucleic acids and to cause free radical formation within the cellular and nuclear milieu. The decrease in nuclear membrane charge and the formation of invaginations are time- and temperature-dependent, occurring maximally at 6 hr and being inhibited at 4°C (Sato et al. 1975). Table 3 indicates that CCNU caused a similar decrease in negative membrane charge when measurements were made 6 hr following drug treatment.

Direct alkylation and carbamoylation of membrane macromolecules may account in part for the change in charge. However, it is more likely that the drug treatment induced a rearrangement of the molecules within the membrane, thereby altering the charge density at the membrane surface.

Although the precise mechanisms by which either X-rays or drugs induce alterations of membrane charge are unknown, the correlation of this phenomenon with cell death has been demonstrated (Sato et al. 1975). The electron micrograph depicting a dying cell (see Fig. 2) is evidence that macromolecular rearrangement and chromatin detachment from the INM accompany drug-induced cytotoxicity. Once again, whether this is the cause of, or a concomitant result of cell death is at present an enigma.

Barrack ER, Coffey DS (1980). The specific binding of estrogens and androgens to the nuclear matrix of sex hormone responsive tissues. J Biol Chem 255:7265.
Barranco SC, Drewinko B, Humphrey RM (1973). Differential response by human melanoma cells to 1,3-bis-(2-chloroethyl)-1-nitrosourea and bleomycin. Mutation Res 19:277.
Biedler JL, Spengler BA (1976). Metaphase chromosome anomaly association with drug resistance and cell specific products. Science 191:185.
Conner GE, Noonan NE, Noonan KD (1980). Nuclear envelope of Chinese hamster ovary cells. Reformation of the nuclear envelope following mitosis. Biochemistry 19:277.
Hemminki K, Vainio H (1979). Preferential binding of benzopyrene into nuclear matrix fraction. Cancer Lett 6:167.
Heppner GH, Dexter DL, DeNucci T, Miller FR, Calabresi P (1978). Heterogeneity in drug sensitivity among tumor

cell subpopulation of a single mammary tumor. Cancer Res 38:3758.
Houghton PJ, Houghton JA, Brodeur GM, Green AA (1981). Characterization, chemosensitivity and drug resistance in xenografts of human childhood rhabdomyosarcomas. Proc Am Assoc Cancer Res 22:245.
Kovacs G (1979). Homogeneously staining regions on marker chromosomes in malignancy. Int J Cancer 23:299.
Maul G (1977). Nuclear pore complexes. Elimination and reconstruction during mitosis. J Cell Biol 74:492.
Mitelman F, Mark J, Levan G, Levan A (1972). Tumor etiology and chromosome pattern. Science 176:1340.
Pratt WB, Ruddon RW (1979). "The Anticancer Drugs." London: Oxford University Press, p 64.
Prestayko AW, Crooke ST, Baker LH, Carter SK, Schein PS (eds) (1981). "Nitrosoureas: Current Status and New Developments", New York: Academic Press.
Price CC (1975). Chemistry of alkylation. In Sartorelli AC, Johns DG (eds): "Antineoplastic and Immunosuppressive Agents", Part II, Berlin: Springer Verlag, p 1.
Sato C, Kojima K, Matsuzawa T, Hinuma Y (1975). Relationship between loss of negative charge on nuclear membrane and loss of colony-forming ability in X-irradiated cells. Radiation Res 62:250.
Shapiro JR, Yung WA, Shapiro WR (1981). Isolation, karyotype and clonal growth of heterogeneous subpopulation of human malignant gliomas. Cancer Res 41:2349.
Szekely JG, Copps TP, Morash BD (1980). Radiation-induced invaginations of the nuclear matrix. Mutation Res 83:621.
Talley RW (1973). Corticosteroids. In Holland JF, Frei E (eds): "Cancer Medicine", Philadelphia: Lea and Febiger, p 1411.
Tew KD, Pinsky SD, Schein PS, Smulson ME, Woolley PV (1980a). Molecular aspects of nitrosamide carcinogenicity. In Davis W, Harrap KR, Stathopolous G (eds): "Human Cancer. Its Characterization and Treatment", Excerpta Medica 5, p 387.
Tew KD, Schein PS, Lindner DJ, Wang AL, Smulson ME (1980b). Influence of hydrocortisone on the binding of nitrosoureas to nuclear chromatin subfractions. Cancer Res 40:3697.
Tew KD, Wang AL, Schein PS (1980c). The nuclear matrix as a target of chloroethylnitrosoureas. Proc Am Assoc Cancer Res 21:297.
Tew KD, Smulson ME, Schein PS (1981). Molecular pharmacology of nitrosoureas. In Carter S, Sakwai Y, Umezawa H

(eds): "Recent Results in Cancer Research", Vol 76, Berlin: Springer Verlag, p 130.

Tew KD, Wang AL, Lindner DJ, Schein PS (1981). Enhancement of nitrosourea cytotoxicity in vitro using hydrocortisone. Submitted for publication.

Tew KD, Wang AL, Schein PS (1981). Chloroethylnitrosourea interactions with the nuclear matrix. Cancer Res, in press.

Wang AL, Schein PS, Tew KD (1981). Nitrosourea interactions with the nuclear envelope, nuclear matrix and associated chromatin. Proc Am Assoc Cancer Res 22:235.

Wang AL, Tew KD (1981). Nitrosourea induced cytotoxicity in a carcinoma resistant to bifunctional alkyating agents. The Pharmacologist 23:154.

Wilkinson R, Birbeck M, Harrap KR (1979). Enhancement of nuclear reactivity of alkylating agents by prednisolone. Cancer Res 39:4256.

THE TOPOLOGY OF DNA LOOPS: A POSSIBLE LINK BETWEEN THE NUCLEAR MATRIX STRUCTURE AND NUCLEIC ACID FUNCTION

Andrew P. Feinberg and Donald S. Coffey

Johns Hopkins Oncology Center
School of Medicine
Baltimore, MD 21205

As the sol-gel image of the nucleus gives way to a more fibrillar-trabecular concept of nuclear structure, the nuclear matrix may play a central role in our understanding of the relationship between DNA structure and function. We will re-examine some traditional aspects of nuclear structure, function, and pathology in light of recent discoveries regarding the nuclear matrix, in order to suggest new perspectives for visualizing the function of eucaryotic DNA.

The nuclear matrix appears to play a central role in the structural organization of both DNA and nuclear RNA. The nuclear matrix contains residual structural elements of the pore complex, lamina, internal ribonucleoprotein (RNP) network and nucleolus (Berezney, Coffey 1977). DNA appears to be attached to the matrix and is organized in supercoiled loops, each containing 30,000-100,000 base pairs, and each loop is anchored at its base to the matrix. During replication, these loops of DNA appear to be reeled through fixed sites for DNA synthesis that are attached to the matrix (Pardoll et al. 1980; Vogelstein et al. 1980). Actively transcribed genes are enriched on the matrix (Pardoll, Vogelstein 1980; Nelkin et al. 1980; Robinson et al. 1982), and it has been suggested that RNA may also be synthesized at fixed sites on the nuclear matrix (Jackson et al. 1981). RNP particles are part of the matrix, and hnRNA and snRNA are associated almost exclusively with the matrix (van Eekelen, van Venrooij 1981; Herman et al. 1978; Berezney, Coffey 1977; Miller et al. 1978a, b). Thus, the nuclear matrix appears to play a central role in DNA replication and transcription and may be involved in RNA processing and transport. The matrix structure also contains specific binding sites for steroid hormones (Barrack, Coffey 1980). For a recent review of nuclear matrix functions, see Barrack, Coffey 1982. We believe that DNA arrangement with respect to the nuclear matrix may be important in many nuclear functions and this may be mediated by DNA loop topology.

THE STRUCTURAL ORGANIZATION OF EUCARYOTIC DNA

In Table 1, we have summarized the various levels of DNA structural organization in the nucleus: a) packing of DNA by histone proteins into nucleosomal units; b) packing of nucleosomes into a chromatin fiber or solenoid; c) structural organization of the fibers into DNA loop domains attached at their base to the matrix; and d) condensation into chromosomes during prophase and metaphase for subsequent separation during anaphase. Overall, the double helix of DNA must be reduced in axial length along the chromosome approximately 8,000-fold (e.g., in each chromatid of human chromosome #1 a DNA molecule that, if completely relaxed would be fully extended to 8 cm, is condensed to about 10 microns). DNA packing by histone proteins to form nucleosomes and linker DNA into 30-nm fibers results in a total packing ratio of about 40.

The next level of complexity is the formation of these fibers into loops that are attached at their base to the non-histone proteins of the nuclear matrix structure. These loops appear to be important in DNA function. A single DNA loop is replicated as a unit and is equivalent to a replicon. In addition, it is interesting that the length of these DNA loops is sufficient to encompass large gene clusters. DNA of the same length as a loop appears to become preferentially sensitive to DNase I during gene activation (Schrader et al. 1981). This may suggest that loop lengths of DNA may be important functional units in both DNA replication and transcription. There appear to be approximately 20,000-50,000 of these loops per nucleus since each loop is approximately 10-30 microns in length and contains 30,000 to 100,000 base pairs (Pardoll et al. 1980; Vogelstein et al. 1980; Razin et al. 1978). The nuclear matrix has been reported to be enriched in repetitive DNA sequences (Razin et al. 1978; Matsumoto, Gerbi 1981; Small et al. 1981), and repetitive DNA may be found at the boundaries of genes and gene clusters. Thus, common sequences may eventually be found in specific areas of the loop and may be associated with their binding to the nuclear matrix.

The final level of structural organization is condensation of the interphase chromatin to form the metaphase chromosome. A metaphase chromosome contains bands which can be further resolved by observation during prophase when they are extended in length up to two-fold (Yunis et al. 1978). Under these conditions, approximately 75 bands can be counted in human chromosome #1. If there are 100 kilobase pairs per loop, there would be about 2000 loops in chromosome #1 or about 30 loops per average high resolution band. An average high resolution band in prophase may thus be condensed in metaphase to occupy an average axial length of only 100-150 nm, which cannot be much wider than the combined diameter of the two chromatin fibers comprising a loop (2 x 30 nm). In order to arrange 30-100 loops in a single band, loops may be positioned radially about the central axis of the chromatid. At present, it is not known what constitutes a chromosomal band, but spatial limitations are imposed by the dimensions of the chromosome, the total content of DNA, the DNA packing ratios and the size and number of the loop domains. The central core of the chromosome has not been fully defined, although residual proteins attached to the base of the loop domains have been visualized (Paulson, Laemmli 1977; Adolph et al. 1977; Stubblefield 1973). However, a controversy exists over the presence and nature of the core structure (Okada, Comings 1980). Some nuclear matrix proteins from the interphase nucleus do appear in the metaphase chromosome (Matsui et al. 1981; Peters et al. 1981). A calculation of the combined nucleosomal volume in a chromosome appears to account for well over 90% of the metaphase chromosome volume, indicating almost complete filling of chromosomal volume by the nucleosomes and leaving very little free space. However, the chromosomal core can be stained with silver (Howell, Hsu 1979; see review by Stubblefield 1981). The formation of nucleosomes, fibrils, and loops produces topological problems for the DNA molecule.

A topological perspective of DNA structure is presented in Table 1 and includes the number of turns or the linking number the DNA molecule undergoes relative to a relaxed double helix. A linking number is defined as the number of times the closed line formed by one edge of a ribbon crosses

Table 1 The Higher-Order Structure of DNA

Structure (unit)	Number of base pairs	Dimensions	Packing ratio[a]	Linking number[b]	Number of nucleosomes	Number of loops	Number of Bands (high resolution)
Double helix (turn)	10.4	3.2 nm pitch 2.0 nm diam.	1	+1	-	-	-
Nucleosome -H1 +H1	146 166	11 x 11 x 5.5 nm	~5	-1.75 -2	1	-	-
Fibril (core + linker)	200	10 nm diam.	~5-7	?	1	-	-
Fiber (turn)	~1,200	10 nm pitch 30 nm diam.	~40	~-6[c]	~6	-	-
Gene (intron, exon)	~1,000- 10,000	?	?	?	~50	-	-
Loop -histones +histones	30,000- 100,000	10-30 μ contour length 0.2-1.0 μ contour length	100-500	(≈) 150-500	0 ~150-500	1	-
Metaphase Band (average)	2-4 x 10⁶	100-150 nm axial length	8,000-10,000	?	~1-2 x 10⁴	30-100	1
Metaphase chromatid	0.5-2.4 x 10⁸	1-10 μ length 0.5 μ diam.	8,000-10,000	?	~0.25-1 x 10⁶	500-2000	15-75
Interphase nucleus	7 x 10⁹	10 μ diam.	?	?	~3.5 x 10⁷	20,000- 50,000	-

[a] The length of DNA in the unit structure if it were fully extended as a relaxed double helix, divided by the axial length of the unit containing the packed DNA.
[b] The number of times one edge of a DNA helix crosses the other edge, relative to the relaxed double helix which is considered as a reference zero (see text).
[c] When the nucleosomes form a chromatin fiber, the linking number is approximately 1 per nucleosome (see text).

the other edge (Crick 1976), and drawings of these stuctures for the DNA molecule have been presented in a popular review (Bauer et al. 1980). The double helix itself contributes +1 (right-handed turns) to the linking number per 10.4 base pairs. The contribution of the double helix of +1 (right-hand) linking number/turn, analogous to the twist of a ribbon, is usually ignored because measurements are made with reference to the relaxed double helix as the reference zero. However, the twist of the double helix may become important if DNA is present in different helical forms, such as Z-DNA (see review by Cantor 1981).

The packing of DNA around the nucleosome contributes about -1.75 to -2.0 to the total linking number since the double helix is wound about the nucleosome 1 3/4 to 2 times, depending upon the presence of the H1 histone. Thus, each nucleosome, containing about 200 (160 core + 40 linker) base pairs would contribute about -2 to the linking number. However, observations of the SV40 minichromosome suggest that every 200 base pairs contributes only -1 to the linking number (Germond et al. 1975). This discrepancy has been called the linking number paradox. To resolve this paradox,

Worcell et al. (1981) have recently proposed that the 30-nm fiber is organized with the nucleosomes in a zigzag pattern in a twisted ribbon arrangement rather than as a simple coil or solenoid. Other topological solutions to this problem also may exist (Gregoryev, Ioffe 1981). This suggests that there may be a more complex order to the loop domain than simple coiling.

Topological considerations of DNA loop structure place severe mechanical constraints on DNA structure that may be important factors in the regulation of DNA function. These constraints may be regulated by specific topoisomerase-like activities, and Vogelstein and Pardoll (1981) have recently reported topoisomerase activity associated with the nuclear matrix, which is consistent with a fixed site of DNA replication on the matrix. Present knowledge of the topology of DNA is in its infancy and even simple topological systems are often difficult to visualize. For example, Figure 1 presents the longitudinal splitting of a paper ribbon previously constructed as a closed loop of linking number 0, 1/2 or 1. Figure 1 demonstrates the surprisingly complex effects of this simple topological operation performed on closed loops of very low linking number. A DNA loop of 100,000 base pairs could possess a linking number of -500 if there were 1 linking number generated per nucleosome (see Table 1). DNA loops attached to the matrix are capable of forming supercoils (Vogelstein et al. 1980). Since loops of DNA form a unit of DNA topology, their attachment to the matrix could provide an important locus for their topological control. Gene activation appears to involve both matrix association (Pardoll, Vogelstein 1980; Nelkin et al. 1980; Robinson et al. 1982) and increased DNase I sensitivity of segments of DNA about the length of loops (see review by Schrader et al. 1981) and changes in internucleosomal configuration (Scheer 1978). Steroid hormones have been reported to increase the sensitivity to DNase I of genes that it induces (Gerber-Huber et al. 1981) and steroid hormones bind specifically to the nuclear matrix (Barrack, Coffey 1980). These topological changes must involve changes in loop shape and supercoiling to account for differences seen in chromatin shape, matrix association, and endonuclease sensitivity. Therefore, the toplogical control of loops at their base attachments to the matrix may play a significant role in gene expression.

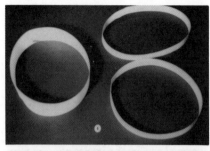

Fig. 1. The result of dividing a closed ribbon of linking number 0, 1/2 or 1. The loop on the left of each photograph is before division; on the right is following division.

DNA LOOPS AND REARRANGEMENT

Just as a loop should not be viewed as a topologically static structure with respect to its shape and supercoiling, a loop also may not be static in its specific location within the genome. Transposable genetic elements, discovered first in maize (McClintock 1950), have been identified in both procaryotes and eucaryotes (Starlinger 1977; Calos, Miller 1980). Some transposable elements play an important role in normal phenotypic expression, such as control of the mating type of yeast (Hicks et al. 1979). Transposable elements have not yet been identified in mammals, although other forms of DNA rearrangement have been seen. The immunoglobulin genes are rearranged by deletion of DNA during lymphocyte maturation (Early et al. 1980; Max et al. 1979; Sakano et al. 1979; Sakano et al. 1980). Transposable elements are generally about gene-size (approximately 1-10 kb), while matrix bound loops may be much larger (30-100 kb). Transposition in procaryotes seems to involve replicon fusion followed by DNA replication (Harshey, Bukhari 1981); in eucaryotes, and loops represent replicon units with DNA synthesis occurring at fixed points attached to the matrix (Pardoll et al. 1980). Transposable elements are characterized by repeat sequences at their ends, and repetitive DNA has been reported to be enriched on the matrix (Matsumoto, Gerbi 1981; Razin et al. 1978; Small et al. 1981). Such repetitive sequences may resemble procaryotic and eucaryotic replication origins (Jelinek et al. 1980). Electron microscopy suggests that procaryotic transposable elements are reeled through a fixed site of replication during transposition (Harshey, Bukhari 1981). In addition, transposition may require topoisomerase-like activity (Liu, Miller 1981), and topoisomerase activity appears to be associated with the matrix (Vogelstein, Pardoll 1981). We therefore propose that loops or segments of loops attached to the matrix might in some cases be capable of rearrangement.

Complexes of many joined looplike structures can be generated from circular duplex DNA by topoisomerases derived from both procaryotes (Kreuzer, Cozzarelli 1980) and eucaryotes (Hsieh, Brutlag 1980). It is thus most intriguing to speculate that eucaryotic loops on a nuclear matrix may have evolved by duplications and transpositions of primordial looplike structures. Much of the so-called "selfish DNA" in eucaryotes (Doolittle, Sapienza 1980; Orgel, Crick 1980) is repetitive and thus an attractive candidate for these loop reorganizations, as discussed earlier. The apparent rearrangement during evolution of genes and gene clusters from one genomic location to another might also be explained by this rearrangement of loops.

DNA PATHOLOGY

One of the hallmarks of cancer pathology is the concomitant development of tumor aggressiveness and changes in cell morphology. The nuclei become enlarged and invaginated and have a pleiomorphic structure which is reflected in both the nucleolus and in changes in the packing of the chromatin. Since the matrix determines the shape of the nucleus, such gross physical changes must involve the nuclear matrix in at least a passive way. In addition, chromosomal rearrangements are seen in a wide variety of cancers, and specific rearrangements appear nonrandomly in some cancers, such as the Philadelphia chromosome in chronic myelogenous leukemia (see review by Sandberg 1980). In addition, most

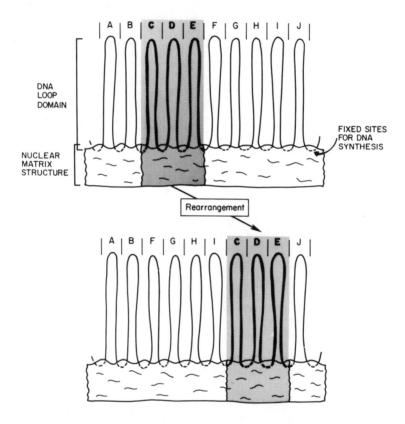

Fig. 2. Chromosomal rearrangements may involve the concomitant movement of the nuclear matrix component to which supercoiled loop domains are attached.

carcinogens cause gross chromosomal aberrations or sister chromatid exchanges.

When gross chromosomal rearrangements of DNA occur in cancer, they would appear to require concomitant involvement of the nuclear matrix component to which the loops are attached (Fig. 2), since the higher order topology of bands appears to be preserved when several bands are translocated as a group from one chromosome to another. It has been reported recently that carcinogens bind to the nuclear matrix (Blazsek et al. 1979; Hemminski, Vainio 1979; Ueyama et al. 1981), and the nuclear matrix also appears to be the site of attachment of the SV40 viral transformation (T) antigen (Buckler-White et al. 1980). Zbarsky (1981) has reported changes in the nuclear matrix in tumor cells. The nuclear matrix contains the fixed sites for DNA replication (Pardoll et al. 1980), and the site of DNA replication may be the locus of sister chromatid exchange (Kato 1980), a form of rearrangement induced by many carcinogens. It is interesting that cells from patients with Bloom's syndrome, which predisposes to both chromosomal rearrangements and cancer, show slow "movement" of the replicating fork (Hand, German 1975), which appears to be located on the nuclear matrix. We do not know if the nuclear matrix is actively involved in carcinogenesis. However, the resemblance of matrix-associated loops to transposable genetic elements noted earlier makes this an intriguing possibility.

In summary, we propose that the nuclear matrix serves as a locus of topological control of DNA loop domains. Specifically, the matrix is involved in the structural organization of the genome, loop replication and gene activation. Since DNA loops are attached at their base to the matrix, this structure could provide an ideal system for hormonal control. Thus the topological arrangement of DNA loops with respect to the nuclear matrix may play an important role in the relationship between nucleic acid structure and function.

ACKNOWLEDGEMENTS

This work was supported by USPHS, NIAMD grant 22000 and USPHS-NCI training grant CA-09071.

Adolph KW, Cheng SM, Paulson JR, Laemmli UK (1977). Isolation of a protein scaffold from mitotic HeLa cell chromosomes. Proc Natl Acad Sci USA 74:4937.

Barrack ER, Coffey DS (1980). Specific binding of estrogens and androgens to the nuclear matrix of sex hormone responsive tissues. J Biol Chem 255:7265.

Barrack ER, Coffey DS (1982). Biological properties of the nuclear matrix: steroid hormone binding. Recent Prog Hormone Res 38, in press.

Bauer WR, Crick FHC, White JH (1980). Supercoiled DNA. Sci Am 243:118.

Berezney R, Coffey DS (1977). Nuclear matrix isolation and characterization of a framework structure from rat liver nuclei. J Cell Biol 73:616.

Blazsek I, Vaukhonen M, Hemminski K (1979). Binding of benzo(a)pyrene into lung and thymocyte nuclear fractions. Res Commun Chem Path Parmacol 23:611.

Buckler-White AJ, Humphrey GW, Pigiet V (1980). Association of polyoma T antigen and DNA with the nuclear matrix from lytically infected 3T6 cells. Cell 22:37.

Calos MP, Miller H (1980). Transposable elements. Cell 20:579.

Cantor CR (1981). DNA choreography. Cell 25:293.

Crick FHC (1976). Linking numbers and nucleosomes. Proc Natl Acad Sci USA 73:2639.

Doolittle WF, Sapienza C (1980). Selfish genes, the phenotype paradigm and genome evolution. Nature 284:601.

Early P, Huang H, Davis M, Calame K, Hood L (1980). An immunoglobulin heavy chain variable region gene is generated from three segments of DNA: V_H, D, and J_H. Cell 19:981.

Gerber-Huber S, Felber BK, Weber R, Ryffel GU (1981). Estrogen induces tissue specific changes in the chromatin conformation of the vitellogenin genes in Xenopus. Nucleic Acids Res 9:2475.

Germond JE, Hirt B, Oudet P, Gross-Bellard M, Chambon P (1975). Folding of the DNA double helix in chromatin-like structures from simian virus 40. Proc Natl Acad Sci USA 72:1843.

Gregoryev SA, Ioffe LB (1981). The dependence of the linking number of a circular minichromosome upon the shape and orientation of its nucleosomes. FEBS Lett 130:43.

Hand R, German J (1975). A retarded rate of DNA chain growth in Bloom's syndrome. Proc Natl Acad Sci USA 72:758.

Harshey RM, Bukhari AI (1981). A mechanism of DNA transposition. Proc Natl Acad Sci USA 78:1090.

Hemminski K, Vainio H (1979). Preferential binding of benzo(a)pyrene into nuclear matrix fraction. Cancer Lett 6:167.
Herman R, Weymouth L, Penman S (1978). Heterogeneous nuclear RNA-protein fibers in chromatin-depleted nuclei. J Cell Biol 78:663.
Hicks JB, Strathern JN, Klar AJS (1979). Transposable mating type genes in Saccharomyces cerevisiae. Nature 282:478.
Howell WM, Hsu TC (1979). Chromosome core structure revealed by silver staining. Chromosoma 73:61.
Hsieh T-s, Brutlag D (1980). ATP-dependent DNA topoisomerase from D. melanogaster reversibly catenates duplex DNA rings. Cell 21:115.
Jackson DA, McCready SJ, Cook PR (1981). RNA is synthesized at the nuclear cage. Nature 292:552.
Jelinek WR, Toomey TP, Leinwand L, Duncan CH, Biro P, Choudary PV, Weissman SM, Rubin CM, Houck CM, Deininger PL, Schmid CW (1980). Ubiquitous, interspersed repeated sequences in mammalian genomes. Proc Natl Acad Sci USA 77:1398.
Kato H (1980). Evidence that the replication point is the site of sister chromatid exchange. Cancer Genet Cytogenet 2:69.
Kreuzer KN, Cozzarelli NR (1980). Formation and resolution of DNA catenanes by DNA gyrase. Cell 20:245.
Liu LF, Miller KG (1981). Eukaryotic DNA topoisomerases: two forms of type I DNA topoisomerases from HeLa cell nuclei. Proc Natl Acad Sci USA 78:3487.
Matsui S, Antoniades G, Basler J, Berezney R, Sandberg AA (1981). Nuclear matrix versus chromosome scaffold: structural basis of eukaryotic chromatin. J Cell Biol 91:60a.
Matsumoto LH, Gerbi SA (1981). The enrichment of satellite DNA on the nuclear matrix of bovine cells. J Cell Biol 91:129a.
Max EE, Seidman JG, Leder P (1979). Sequences of five potential recombination sites encoded close to an immunoglobulin κ constant region gene. Proc Natl Acad Sci USA 76:3450.
McClintock B (1950). The origin and behavior of mutable loci in maize. Proc Natl Acad Sci USA 36:344.
Miller TE, Huang C-Y, Pogo AO (1978a). Rat liver nuclear skeleton and ribonucleoprotein complexes containing hnRNA. J Cell Biol 76:675.

Miller TE, Huang C-Y, Pogo AO (1978b). Rat liver nuclear skeleton and small molecular weight RNA species. J Cell Biol 76:692.

Nelkin BD, Pardoll DM, Bogelstein B (1980). Localization of SV40 genes within supercoiled loop domains. Nucleic Acids Res 8:5623.

Okada TA, Comings DA (1980). A search for protein cores in chromosomes: is the scaffold an artifact? Am J Hum Genet 32:814.

Orgel LE, Crick FHC (1980). Selfish DNA: the ultimate parasite. Nature 284:604.

Pardoll DM, Vogelstein B (1980). Sequence analysis of nuclear matrix associated DNA from rat liver. Exp Cell Res 128:466.

Pardoll DM, Vogelstein B, Coffey DS (1980). A fixed site of DNA replication in eucaryotic cells. Cell 19:527.

Paulson JR, Laemmli UK (1977). The structure of histone-depleted metaphase chromosomes. Cell 12:817.

Peters KE, Okada TA, Comings DE (1981). 2D gel electrophoresis of interphase and metaphase nonhistone proteins. J Cell Biol 91:72a.

Razin SV, Mantieva VL, Georgiev GP (1978). DNA adjacent to attachment points of deoxyribonucleoprotein fibril to chromosomal axial structure is enriched in reiterated base sequences. Nucleic Acids Res 5:4737.

Robinson S, Nelkin B, Vogelstein B (1982). Ovalbumin gene is associated with the nuclear matrix of hen oviduct cells. Cell, in press.

Sakano H, Huppi K, Heinrich G, Tonegawa S (1979). Sequences at the somatic recombination sites of immunoglobulin light-chain genes. Nature 280:288.

Sakano H, Maki R, Kurosawa Y, Roeder W, Tonegawa S (1980). Two types of somatic recombination are necessary for the generation of complete immunoglobulin heavy-chain genes. Nature 286:676.

Sandberg AA (1980). "The Chromosomes in Human Cancer and Leukemia". New York: Elsevier North Holland.

Scheer U (1978). Changes of nucleosome frequency in nucleolar and non-nucleolar chromatin as a function of transcription: an electron microscopic study. Cell 13:535.

Schrader WT, Birnbaumer ME, Hughes MR, Weigel NL, Grody WW, O'Malley BW (1981). Studies on the structure and function of the chicken progesterone receptor. Recent Prog Hormone Res 37:583.

Small D, Nelkin B, Vogelstein B (1981). Nonrandom association of repeated DNA sequences with the nuclear matrix. Submitted.

Starlinger P (1977). DNA rearrangements in procaryotes. Ann Rev Genet 11:103.

Stubblefield E (1981). The molecular organization of mammalian metaphase chromosomes. In Arrighi FE, Rao PN, Stubblefield E (eds): "Genes, Chromosomes, and Neoplasia", New York: Raven Press, p 61.

Stubblefield E (1973). The structure of mammalian chromosomes. Int Rev Cytol 35:1.

Ueyama H, Matsuura T, Nomi S, Nakayasu H, Ueda K (1981). Binding of benzo(a)pyrene to rat liver nuclear matrix. Life Sci 29:655.

van Eekelen CAG, van Venrooij WJ (1981). hnRNA and its attachment to a nuclear protein matrix. J Cell Biol 88:554.

Vogelstein B, Pardoll DM (1981). Direct visualization of topoisomerase activity in subnuclear structures. Submitted.

Vogelstein B, Pardoll DM, Coffey DS (1980). Supercoiled loops and eucaryotic DNA replication. Cell 22:79.

Worcell A, Strogats S, Riley D (1981). Structure of chromatin and the linking number of DNA. Proc Natl Acad Sci USA 78:1461.

Yunis JJ, Sawyer JR, Ball DW (1978). The characterization of high-resolution G-banded chromosomes of man. Chromosoma 67:293.

Zbarsky IB (1981). Nuclear skeleton structures in some normal and tumor cells. Mol Biol Rep 7:139.

Small D, Nelkin B, Vogelstein B (1981). Nonrandom association of repeated DNA sequences with the nuclear matrix. Submitted.

Starlinger P (1977). DNA rearrangements in procaryotes. Ann Rev Genet 11:103.

Stubblefield E (1981). The molecular organization of mammalian metaphase chromosomes. In Arrighi FE, Rao PN, Stubblefield E (eds): "Genes, Chromosomes, and Neoplasia", New York: Raven Press, p 61.

Stubblefield E (1973). The structure of mammalian chromosomes. Int Rev Cytol 35:1.

Ueyama H, Matsuura T, Nomi S, Nakayasu H, Ueda K (1981). Binding of benzo(a)pyrene to rat liver nuclear matrix. Life Sci 29:655.

van Eekelen CAG, van Venrooij WJ (1981). hnRNA and its attachment to a nuclear protein matrix. J Cell Biol 88:554.

Vogelstein B, Pardoll DM (1981). Direct visualization of topoisomerase activity in subnuclear structures. Submitted.

Vogelstein B, Pardoll DM, Coffey DS (1980). Supercoiled loops and eucaryotic DNA replication. Cell 22:79.

Worcell A, Strogats S, Riley D (1981). Structure of chromatin and the linking number of DNA. Proc Natl Acad Sci USA 78:1461.

Yunis JJ, Sawyer JR, Ball DW (1978). The characterization of high-resolution G-banded chromosomes of man. Chromosoma 67:293.

Zbarsky IB (1981). Nuclear skeleton structures in some normal and tumor cells. Mol Biol Rep 7:139.

EVOLUTION OF THE NUCLEAR MATRIX AND ENVELOPE

T. Cavalier-Smith

Department of Biophysics
King's College London
26-29 Drury Lane
London WC2B 5RL

This paper is concerned with the evolution of the nuclear matrix and nuclear envelope, and must consist largely of unanswered questions that may help to stimulate future research. The origin of the nuclear envelope has been discussed before (Cavalier-Smith 1975, 1980a, 1981a; Maul 1977) but the origin of the nuclear matrix has not, despite growing evidence for its importance in nuclear architecture and function (see Kaufmann et al. 1981; Agutter, Richardson 1980; Jost, Johnson 1981, and the other articles in this volume for reviews).

The nuclear matrix consists of two parts: (1) the nuclear lamina underlying the nuclear envelope, to which the nuclear pore complexes and the inner membrane of the envelope are attached, and (2) an inner nuclear matrix that lies inside the nuclear lamina and fills that part of the nucleus not occupied by the chromatin. For brevity, I shall refer to these as the lamina and the inner matrix. In interphase nuclei, the DNA of chromatin is strongly bound both to the lamina and to the inner matrix, but both components of the matrix retain their integrity after the removal of the DNA by DNase and the removal of the nuclear envelope by non-ionic detergents. Though the lamina and inner matrix may have some structural proteins in common, they differ in that the structural integrity of the lamina is independent of RNA whereas that of the inner matrix is apparently disrupted if RNA is removed under conditions where cross-listing by disulphide bonds is prevented (Kaufmann et al. 1981). In mammalian cells the major lamina proteins, lamin A, B and C, are pre-dominantly located in the lamina in interphase but are widely dispersed throughout the cell during mitosis following the breakdown of the nuclear envelope.

IS THE NUCLEAR MATRIX PRESENT IN ALL EUKARYOTES?

The above-mentioned properties of the matrix are based entirely on studies of vertebrates, mainly mammals. A nuclear matrix has also been isolated from the macronucleus of the protozoan Tetrahymena (Wunderlich, Herlan 1977), but it is not yet clear how similar it is to the vertebrate nuclear matrix.

There is a great need to study fungi, protozoa, algae and higher plants to see whether or not they possess nuclear matrices, and if so, how similar they are in the different groups. Since the nuclear envelope and pore complexes have a remarkably uniform structure in all eukaryotes (Franke 1970; Maul 1977) and are intimately associated with the nuclear lamina in mammals, the simplest assumption, which I shall adopt here, is that the lamina at least is present in all eukaryotes, though its detailed properties are likely to vary. In view of the evidence that DNA replication forks are associated with the nuclear matrix (Berezney, Buchholtz 1981), and the likelihood that the basic replication pattern for DNA is conserved throughout eukaryotes, I shall also assume that some kind of ribonucleoprotein inner matrix is present in all eukaryotes. In protists as diverse as yeasts and dinoflagellates, enzymatic removal of DNA leaves a considerable intranuclear residue (Gordon 1977). However, it is not yet known if it remains when detergent is used to remove the envelope as well.

THE NUCLEAR MATRIX AND CLOSED MITOSIS

If the matrix is universally present in eukaryotes, the next fundamental question is how does it behave during mitosis in the numerous protists in which the nuclear envelope remains intact throughout cell division.

If such organisms have a lamina it must, during mitosis, either lose its rigidity or disperse completely as in mammals. One indication that the nuclear matrix might be able

to disperse even in the absence of nuclear envelope breakdown is the existence of a variety of protists with closed or semi-closed mitosis in which the nucleolus disperses during division (Heath 1980). Since in mammals the nucleolus is apparently attached to the matrix (Franke et al. 1981) it is reasonable to suppose that the dispersal of the nucleolus at the end of prophase and its reformation during telophase is characteristic of animals and other organisms where an open mitosis is simply the necessary result of the prophase disassembly and telophase reassembly of the nuclear matrix, or a special component of it.

However, there are also many protists with a closed mitosis in which the nucleolus does not disperse during mitosis but is simply pulled into two halves (Heath 1980). This implies that in these organisms a complete mitotic dispersal of the inner matrix may not occur. It will be important to compare the properties of the matrix throughout the cell cycle in three kinds of organisms: (1) species like Saccharomyces cerevisiae that have a persistent nucelolus and a closed mitosis, (2) those like Aspergillus that show nucleolar dispersal despite having a closed mitosis and, (3) those like animals where both nucleolus and nuclear envelope disperse. This difference may be one of the most important events in the evolution of the nuclear matrix since its origin.

THE ORIGIN OF THE NUCLEAR MATRIX

Discussion of the origin of an organelle can be more realistic if we have some idea of what kind of cell it originated from. According to traditional views of cell evolution, the first eukaryote was either a protozoan or an alga. I have argued in detail elsewhere (Cavalier-Smith 1980a, 1981a, 1981b, 1982a, 1982b) that it is more probable that the first eukaryote was a fungus, similar to modern Hemiascomycetes. Initially, I favored a filamentous mode of growth (Cavalier-Smith 1980a, 1981a, 1982a), but I now think it more probable that the first eukaryote was a budding yeast very similar to Saccharomyces cerevisiae. I also think that it evolved from a purple non-sulphur budding bacterium like Rhodomicrobium or Rhodopseudomonas whose cytochrome c molecules are more similar to those of eukaryotes than to those of any other bacteria (Dayhoff, Schwartz 1981). The purple budding bacteria also contain flattened internal membranes that bear the cytochromes and other molecules of the respiratory chain, which in Rhodomicrobium are arranged around the bacterial DNA rather like a nuclear envelope and which could have lost their photosynthetic pigments and undergone fusion to form mitochondrial and nuclear envelopes as described elsewhere (Cavalier-Smith 1980a, 1981a). This fusion, probably like the prior evolution of histones (Cavalier-Smith 1981a), would serve to protect the DNA from shearing damage by microtubule- and actin-based cytoplasmic motility, which I suggest originated together with exocytosis to speed up cell wall secretion (Cavalier-Smith 1980a, 1981a). In reference to this hypothesis, what triggered the major genetic revolution that gave rise to the eukaryotic cell was the mutational loss of muramic acid from the prokaryotic peptidoglycan cell wall. Its replacement, formed by a chitin/mannan/glucan cell wall, was secreted by the novel process of exocytosis.

There are two reasons for thinking that the nuclear matrix might have originated during this drastic cell reorganization rather than at some later stage. One stems from the fact that, at least in higher eukaryotes, the nuclear envelope membranes appear to be attached to the DNA not directly but through the intermediary of the nuclear lamina. If early eukaryotes lacked a nuclear lamina, then the nuclear envelope must either not have been attached to the DNA at all, which would make a specific association between the two structures difficult, or directly attached to the DNA. In the latter case, the changeover from direct attachment to attachment via lamina proteins is not easy to visualize. The second reason is the involvement of the matrix in DNA replication (Pardoll et al. 1980): If the replication machinery for the numerous separate origins of replication is attached to the nuclear matrix, it is simplest to suppose that this occurred at the early stage in eukaryote evolution when replication forks first became free of the plasma membrane. This feeding of replication forks from the plasma membrane must have been completed by the time the nuclear envelope itself was fully formed. However, this could not have occurred prior to the origin of a primitive mitotic spindle able to segregate chromosomes in place of the prokaryotic mechanism that is thought to depend on the growth in the cell surface (Cavalier-Smith 1980a). This means that the matrix must have evolved at some time during the origin of the basic nuclear structures.

The matrix may therefore be thought of as a substitute for the DNA attachment sites on the prokaryotic plasma membrane. Since the latter are probably specific proteins, they could have been the ancestors of some of the matrix proteins and it will be important to seek homologies between them. The bacterial DNA-attachment proteins must be DNA-binding proteins attached to or embedded in the membrane, as the eukaryotic lamina proteins also must be.

The inner matrix proteins, however, are not membrane-associated proteins, though some of them must be DNA-binding proteins. At some stage, therefore, there must have been an evolutionary differentiation between the lamina and inner matrix. It is possible that this difference will be traceable back to differences in their ancestral proteins in bacteria. Many features of the control of bacterial DNA replication, especially the evidence and logical requirements for differences in the mechanisms of initiation, replication fork movement and termination (Davern 1979), are easiest to understand if bacteria have three different classes of membrane-attached DNA-binding proteins: (1) those that bind replicon origins, (2) those that bind replication forks and/or the replisomes, and (3) those that bind termini. I suggest that the inner matrix evolved from replisome binding proteins by the loss of their association with membrane but that the lamina proteins evolved instead from origin-and/or terminus-binding proteins that retained their association with membranes throughout.

ORIGIN AND FUNCTIONS OF THE NUCLEAR ENVELOPE

The octagonal nuclear pore complexes are firmly attached to the nuclear lamina, and Maul (1977) has suggested that they might have evolved from prokaryotic membrane-associated replicon origin proteins. Another suggestion (Cavalier-Smith 1980a) is that they are derived from membrane-bound ATPases. In our present state of ignorance it may be unproductive to speculate further on their possible precursors, except perhaps to emphasize the possibility of their constituent polypeptides having had a mosaic origin as a result of the fusion of segments from a variety of prokaryotic genes. There are several reasons for thinking that DNA-transpositions played a major role in the origin of the eukaryotic cells (Cavalier-Smith 1980a, 1981a).

An important basic point about the origin of the nuclear envelope, in marked contrast to the mitochondrial envelope, is that, at no stage, can there have been a complete fusion of membranes that would seal the nuclear envelope off from the cytoplasm and prevent the exit of RNA and the entry of proteins. This suggests that the nuclear pore complex, or something functionally equivalent, must have been there from the start. Indeed, it may well be that the essential components of the nucleus, and those that evolved first, are the matrix and the pore complexes; the nuclear membranes may perhaps have been added as an 'afterthought'.

It is a curious fact about such a universal organelle that it is not at all clear what function the nuclear envelope membranes have, though there is little doubt concerning the transport functions of the pore complexes (Paine, Horowitz 1980). Unlike the mitochondrial and plastid envelope membranes, the nuclear membranes appear to be unimportant in cellular compartmentation of low molecular weight metabolites like nucleotides (Paine, Horowitz 1980). One possible role for the nuclear membranes is preventing mature ribosomal subunits from entering the nucleus where translation of premessenger RNA from split genes would produce faulty proteins (Cavalier-Smith 1981a). However, the presence of split genes in a few mitochondrial proteins of the yeast Saccharomyces cerevisiae (Mahler 1981) where pre-mRNA and ribosomes are present in the same compartment, suggests that other mechanisms can prevent the translation of pre-mRNA. I therefore suggest that the fundamental role of the nuclear membranes is to act as the cell's basic reservoir of endoplasmic reticulum membranes. If during development new membranes can only arise by the growth and subdivision of pre-existing membranes in accordance with the principle omnis membranis e membranis, then it is essential to preserve a store of membrane throughout the life cycle and to ensure its segregation to both daughters at cell division. In prokaryotes this is achieved by the persistence and division of the plasma membrane from which most, perhaps all, prokaryotic intracellular membranes can be developmentally derived. In eukaryotes, however, there appears to be a fundamental distinction [probably having arisen during the origin of the eukaryotic cell (Cavalier-Smith 1981a)] between the plasma membrane which never bears ribosomes and the endoplasmic reticulum/outer nuclear envelope membrane system which usually does have ribosomes on its cytosolic surface. Although endoplasmic reticulum membrane can move to the cell surface by an indirect route involving Golgi membranes, there is no evidence for the developmental derivation of rough endoplasmic reticulum (RER) membranes from the plasma membrane even by an indirect route. I suggest that RER membranes can arise only by the division of pre-existing RER membranes (I include the outer membrane of the nuclear envelope as part of the RER), and that a fundamental reason for the presence of a nuclear envelope in all eukaryotes is to ensure the segregation of RER membranes to both daughter cells. The persistence of the nuclear envelope during nuclear division in closed mitoses may therefore be not merely a primitive evolutionary relic, but a simple way of ensuring that both daughter cells get some RER. The embedding of the centrosomal plaques in the nuclear envelope as in Saccharomyces, or their attachment to its outer surface as in many other organisms, is consistent with the idea that segregation of RER to both daughters is as vital as segregation of chromosomes.

In open mitosis the nuclear envelope does not disappear as is sometimes wrongly thought. Instead, as Hepler and others have shown (Hepler 1977, 1980), it becomes partially fragmented and spatially rearranged so as to be associated mainly with the spindle poles and chromosomal spindle fibers. Thus, even open mitosis will cause the segregation of a basic store of RER membranes to both daughter cells. Nuclear envelope segregation need not be as precise as chromosomal segregation, but as in the case of plasma membrane segregation at cytokinesis, division in most instances is probably quite accurately equal. It is well known that unequal cytokinesis occurs in certain stages of development to produce daughter cells of unequal volume. It is less widely known that in some tissues unequal nuclear division occurs to produce daughter nuclei of differing volume. Since unequal nuclear division is best documented in organisms having open mitosis, like vascular plants (Gunning et al. 1978), it seems probable that it must depend on the unequal reassembly of nuclear matrix proteins at telophase: that is, more matrix proteins must be reassembled around one set of daughter chromosomes than another. A possible basis for such unequal assembly might be an unequal rate of unfolding of the two daughter chromosome sets. Since many matrix proteins are DNA-binding proteins (Comings, Wallach 1978), the daughter chromosome set that unfolded faster would be able to sequester more of them to produce a larger nuclear size.

Another possible function for nuclear envelope membranes is the control of intranuclear calcium concentrations by pumping calcium into the perinuclear cisterna. Hepler (1977, 1980) has provided evidence and arguments for the view that the rearrangement of nuclear envelope membranes during open mitosis is related to the control of spindle microtubule behavior by the sequestering of calcium. It has long been a puzzle why some species have open mitosis and others have closed. It seems that in general those with large nuclei have open mitosis and those with small nuclei have closed mitosis and it has been proposed that open mitosis is for some reason more suited to larger cells and closed mitosis to smaller cells (Cavalier-Smith 1980b). Hepler's arguments make it possible to see why this should be. I suggest that accurate temporal control of mitosis by calcium would only be possible if the diffusion path between the membranes that sequester or release it is kept short. In a small nucleus, all the microtubules are close to the nuclear envelope so there is no need to reorganize its membranes during mitosis to achieve this; small nuclei will therefore have closed mitosis. In large nuclei it is necessary to reorganize the membranes by partial breakdown of the nuclear envelope to ensure that no microtubules are far removed from them, so large nuclei will have open mitosis.

Two apparent exceptions to the rule that large nuclei have open mitosis are in fact consistent with this idea. One is the dinoflagellates and parabasalian flagellates which can have exceedingly large nuclei yet invariably have closed mitosis. However, they differ from all other eukaryotes having closed mitosis in that the spindle microtubules are all outside the nuclear envelope (Heath 1980); they are therefore quite close to it and the diffusion pathways for calcium would be short even in a large nucleus. Their peculiar mitosis could have arisen by a mutation that caused the kinetochores to insert themselves into a nuclear pore. The second apparent exception is the ciliate macronucleus. Here, however, the arrangement and behavior of the chromatin during mitosis is atypical and the vast majority of the microtubules are very close to the nuclear surface (Tucker et al. 1980). Open mitosis appears to have originated many times independently in different phylogenetic lines (Heath 1980) that have undergone increases in cell and nuclear volume. For example, the isolated occurrence of open mitosis in Basidiobolus, when other zygomycetes have closed mitosis (Heath 1980), strongly suggests that it acquired this property independently of other eukaryotes because it has an exceptionally large nucleus for a zygomycete.

THE NUCLEAR MATRIX AND NUCLEAR VOLUME CONTROL

Perhaps the most frequent evolutionary changes affecting the nuclear matrix are increases or decreases in nuclear volume. Nuclear volumes vary more than a million-fold in different species and tissues. For uninucleated cells there is a strong positive correlation between nuclear and cell volume when species of widely differing cell volume are compared (Strasburger 1883; Olmo, Morescalchi 1975, 1976; Cavalier-Smith 1978, 1982c), so it appears that nuclear volume must vary in evolution approximately (though not exactly) in step with cell volume (Cavalier-Smith 1978, 1980c, 1982c). Because nuclear DNA content also shows a similar variation and correlates strongly with both nuclear and cell volume, I have proposed that it is the nuclear DNA content plus its pattern of folding that together primarily determine the volume of nuclei (Cavalier-Smith 1978, 1982c). This skeletal DNA hypothesis supposes that the nuclear matrix can in principle be assembled at many different volumes depending simply on the geometric arrangement of the DNA that provides nucleating sites for its assembly. Evolutionary changes in the amount or folding pattern of the DNA could therefore change the volume of the nuclear matrix without altering the matrix genes themselves so long as there is a negative feedback mechanism to adjust the amount of matrix proteins synthesized in relation to the number of binding sites on the DNA. An alternative hypothesis would be that the nuclear matrix has an inherent size-determining capacity independent of the DNA and that it is evolutionary changes in this that cause evolutionary changes in nuclear volume.

It is an important goal for nuclear matrix research to determine which of these two hypotheses is correct, both because of the inherent interest in understanding the mechanism of control of nuclear volume - a much neglected topic in cell biology - and because of the importance of understanding the C-value paradox, for which the skeletal DNA hypothesis is one possible solution (Cavalier-Smith 1978, 1980c, 1981c).

ACKNOWLEDGMENTS

Some of the ideas expressed here were developed during the tenure of a Visiting Fellowship at the Australian National University. I thank colleagues there in the Research School of Biological Sciences for stimulating discussions and the Royal Society for a travel grant.

Agutter PS, Richardson CW (1980). Nuclear non-chromatin proteinaceous structures: their role in the organization and function of the interphase nucleus. J Cell Sci 44:395.

Berezney R, Buchholtz LA (1981). Dynamic association of replicating DNA fragments with the nuclear matrix of regenerating liver. Exp Cell Res 132:1.

Cavalier-Smith T (1975). The origin of nuclei and of eukaryotic cells. Nature 250:463.

Cavalier-Smith T (1978). Nuclear volume control by nucleoskeletal DNA, selection for cell volume and cell growth rate, and the solution of the DNA c-value paradox. J Cell Sci 34:247.

Cavalier-Smith T (1980a). Cell compartmentation and the origin of eukaryote membraneous organelles. In Schwemmler W, Schenk HEA (eds): "Endocytobiology: Endosymbiosis and Cell Biology", Berlin: de Gruyter, p 831.

Cavalier-Smith T (1980b). r- and k-Tactics in the evolution of protist developmental systems: cell and genome size, phenotype diversifying selection and cell cycle pattern. Biosynthesis 12:43.

Cavalier-Smith T (1980c). How selfish is DNA? Nature 285:617.

Cavalier-Smith T (1981a). The origin and early evolution of the eukaryotic cell. Symp Soc Gen Microbiol 32:33.

Cavalier-Smith T (1981b). Eukaryote kingdoms: 9 or 7. Biosystems, in press.

Cavalier-Smith T (1982a). The evolutionary origin and phylogeny of eukaryote flagella. Symp Soc Exp Biol, in press.

Cavalier-Smith T (1982b). The origin of mitosis (in preparation).

Cavalier-Smith T (1982c). Skeletal DNA and the evolution of genome size. Ann Rev Bioph Bioeng 11, in press.

Comings DE, Wallach AS (1978). DNA-binding properties of nuclear matrix proteins. J Cell Sci 34:233.

Davern CI (1979). Replication of the prokaryotic chromosome with emphasis on the bacterial chromosome replication in relation to the cell cycle. In Prescott DM, Goldstein L (eds): "Cell Biology: A Comprehensive Treatise", Vol 2, New York: Academic Press, p 131.

Dayhoff MO, Schwartz RM (1981). Evidence on the origin of eukaryotic mitochondria from protein and nucleic acid sequences. Ann NY Acad Sci 361:92.

Franke WW (1970). On the universality of nuclear pore complex structure. Z Zellforsch 105:405.

Franke WW, Kleinschmidt JA, Spring H, Krohne G, Grund C, Trendelenburg MF, Stoehr M, Scheer U. (1981). A nucleolar skeleton of protein filaments demonstrated in amplified nucleoli of Xenopus laevis. J Cell Biol 90:289.

Gordon CN (1977). Chromatin behavior during the mitotic cell cycle of Saccharomyces cerevisiae. J Cell Sci 24:81.

Gunning BES, Hughes JE, Hardham AR (1978). Formative and proliferative cell divisions, cell differentiation and developmental changes in the meristem of Azolla roots. Planta 143:121.

Heath IB (1980). Variant mitoses in lower eukaryotes: indicators of the evolution of mitosis. Int Rev Cytol 64:1.

Hepler PK (1977). Membranes in the spindle apparatus: their possible role in the control of microtubule assembly. In Rost TL, Gifford EM (eds): "Mechanisms and Control of Cell Division", Stroudsburg, Pennsylvania: Dowden, Hutchinson and Ross, p 212.

Hepler PK (1980). Membranes in the mitotic apparatus of barley cells. J Cell Biol 86:490.

Jost E, Johnson RT (1981). Nuclear lamina assembly, synthesis and disaggregation during the cell cycle in synchronized HeLa cells. J Cell Sci 47:25.

Kaufmann SH, Coffey DS, Shaper JH (1981). Considerations in the isolation of rat liver nuclear matrix, nuclear envelope and pore complex lamina. Exp. Cell Res 132:105.

Mahler MR (1981). Mitochondrial evolution: organization and regulation of mitochondrial genes. Ann NY Acad Sci 361:53.

Maul GG (1977). The nuclear and cytoplasmic pore complex: structure, dynamics, distribution and evolution. Int Rev Cytol Suppl 6:75.

Olmo E, Morescalchi A (1975). Evolution of the genome and cell sizes in salamanders. Experientia 31:804.

Olmo E, Morescalchi A (1978). Genome and cell sizes in frogs: a comparison with salamanders. Experientia 34:44.

Paine PL, Horowitz SB (1980). The movement of material between nucleus and cytoplasm. In Prescott DM, Goldstein

L (eds): "Cell Biology: A Comprehensive Treatise", Vol 4, New York: Academic Press, p 299.
Pardoll, DM, Vogelstein B, Coffey DS (1980). A fixed site of DNA replication in eukaryotic cells. Cell 19:527.
Strasburger E (1893). Ueber die Wirkungsphare der Kerne und die Zellgrosse. Histologische Beit 5:97.
Tucker JB, Beisson J, Roche DLJ, Cohen J (1980). Microtubules and control of macronuclear 'amitosis' in Paramecium. J Cell Sci 44:135.
Vogelstein B, Pardoll DM, Coffey DS (1980). Supercoiled loops and eukaryotic DNA replication. Cell 22:79.
Wunderlich F, Herlan G (1977). A reversibly contractile nuclear matrix. J Cell Biol 73:271.

Index

α-Polymerase, 193. See DNA.
A, Medium, contrast with B, C in RNA release, 98. See B, C, mRNA
cAMP
    cell-free system component, 123
    tryptophan with, 123. See Tryptophan.
Antibodies
    ATP and dependent RNA release with, 130–133
    lamina pore complex and, 132, 133. See Lamina and Nuclear Pores.
T-Antigen
    diester linkage phosphate with, 220
    electropheresis acquisition of, 214
    nuclear matrix with, 214, 216. See Nuclear Matrix.
    phosphorylated, 216–20
ATP
    antibodies, 130. See Antibodies.
    cell-free system and RNA transport, 130. See Cell-free System, DNA, Nuclear Envelope, RNA transport.
Autoradiography. See Nuclear Matrix.

B, Medium
    RNA release, 98. See A, C, mRNA.
Bγ bond. See RNA transport.

C-group proteins. See hnRNA.
C, Medium
    contrast with A, B in RNA. See A, B, mRNA.
Cell Death
    nuclear-reactant drugs and, 281. See Nuclear-Reactant Drugs.
    transition and degeneration observed, 280ff. See Nuclear-Reactant Drugs.
Cell-Free System. See cAMP, ATP.
Cell Matrix. See Nuclear Matrix.

Chaotropic Reagents. See Nuclear Pore Complexes.
Chromatin. See Steroid Hormone Action.
Corticosteroid. See Nuclear Matrix.
Crosslinking
    nuclear matrix, 223–32. See Nuclear Matrix.
    nuclear morphology preservation and, 223. See Nuclear Morphology.
C_2. See RNA Efflux.
Cytosol
    RNA release and, 98. See RNA.

DNA. See Nuclear Matrix, Polyoma, Proteins.
    α-polymerase, activity as major enzyme, 193
    ATP effect, 100. See ATP.
    binding, 202
    endogenous endonucleases and, 101
    eucaryotic function, 293. See Eucaryotic Cells.
    structural organization, 294ff.
    functional replication complexes, 193
    lamina and internal matrix, 204
    "leakage," nuclear morphology in RNA efflux, 103, 104. See RNA Efflux.
    loops. See Nuclear Matrix.
    dynamic nature, 183–89
    eucaryotic evolution, 299
    internuclear arrangement, 189
    nuclear matrix and, 294, 297–99
    nuclear matrix and nucleic acid link, 293–301
    supercoiled, 169ff. See Nuclear Matrix.
    characteristics, 172, 173
    mechanical constraints on structure, 297

nuclear stability in RNA efflux, 100. See RNA Efflux.
packing and linking, 296-7
pathology, 299-301
Physarum, 205ff. See *Physarum.*
pulse-chase experiments with, 177, 178. See Nuclear Matrix.
ratios of newly synthesized to bulk, 206
replication and transcription, 1
associated with nuclear matrix, 169-78, 199-210
"sticking," 173ff.
synthesis in nuclear matrix, 193-5. See Nuclear Matrix.
DNAase 1. See *Drosophilia Melanogaster.*
Drosophilia Melanogaster
antibody production and characterization to matrix polypeptides, 148ff.
chemical composition
matrix polypeptide, 150ff.
SDS-Page, 147
glycoprotein content, 150
immunochemical characterization, 149, 151-152. See Nuclear Pore Complex and Lamina.
nuclear structure and function, 147, 151-2
isolated nuclear matrix function, 145-52
structural and biochemical properties, 146ff.
subcellular fractions, 147. See DNAase 1, RNAase A, Triton X-100, and Im NaCl.
extraction conditions, 148
morphological and biochemical integrity, 148

Erythrocytes. See *Xenopus laevis.*
Estrogen
binding sites
components of biosynthetic growth, 266
conversion to tighter binding, 266
nuclear matrix, 259-67. See Nuclear Matrix.

type I & II sites, 266
fractions after beeswax implant, 264-266
salt extraction effects, 261ff.
type II increases, 263. See KCl-extraction, Nuclear Matrix.
receptors
acceptor and non-acceptor association, 259, 260
measurement errors, 264-66
Eucaryotic Cells. See DNA, Nuclear Pores.
components, 1
structural domains, 145. See *Drosophilia Melanogaster.*

H3. See Nonhistone Protein.
HeLa S_3 Cells. See Nuclear Envelope Reformation, Nuclear Matrix.
Histones
nuclear envelopes, 19. See Nuclear Envelope, Nuclear Envelope Fractions, Proteins, Nonhistone

Immunological Probes
RNA transport in nuclear envelope, 129-133. See RNA Transport in Nuclear Envelope.
cell-free, 130
Immunoglobulin. See Nuclear Envelope, RNA Transport.
Im Na Cl. See *Drosophilia Melanogaster.*
Insulin
RNA transport, 89, 90

KCl-extraction. See Estrogen type II binding.

Lamina. See Antibodies, DNA, Nuclear Envelope Fractions, Steroid Hormone Action.
Lamins. See Nuclear Lamins.
A, B, C. See Nuclear Periphery.

Medium A. See RNA Release, B, C.

Mg^{2+}. See Nuclear Matrix, Mg^{++}.
    activating NTPase in RNA efflux, 102. See NTPase, RNA Efflux.
    stabilization and nuclear restriction, 102
Mitotic Cells. See Nuclear Matrix.
    late
        labeling in nuclear envelope reformation, 47. See Nuclear Envelope Formation.
        phospholipid biosynthetic activity, 50. See Phospholipid Metabolism.
$M\gamma$ 46,000 A. See Protein, Nonhistone.
$M\gamma$ 68,000, $M\gamma$ 145,000. See Oocytes, Polypeptides.

N-Proteins. See Nuclear Proteins.
NTPase. See Hepatic Nuclear Envelope, Mg^{2+}, RNA Efflux.
Nuclear Binding, 68-71. See Nuclear Extracts.
Nuclear Envelope. See Nuclear Proteins, Histones, Phospholipid Inhibitors and Metabolism.
    evolution, 307-15
    formation. See Late Mitotic Cells.
    fractions
        polypeptide composition. See Histones and Lamina.
        histones, 19
        lamina, 13-27
            polyvinylsulfate exposure, 20, 21
            solubilization, 25
            urea treatment and analysis, 25, 27
    hepatic, 118
        NTPase with, 118-20. See NTPase.
    internal components, 17
    origins and functions, 311-15
    penetration of, 71
        mechanisms for uptake of polypeptides, 71-72
    permeability, 64
        exogenous and endogenous exchange, 68, 72-73
        nucleocytoplasmic distribution, 65

pore complex lamina
    characteristics and function, 42, 43
    movement between membrane system and pore complexes, 32ff. See Nuclear Membrane and Pore Complex.
    peptide mapping, 160. See Nuclear Pores.
    properties and function, 157-66
    structural associations, 161ff.
    subtypes, 160
    topography, 164, 165
Nuclear Matrix. See T Antigen, DNA, Estrogen.
    autoradiography and biochemical experiments, 172ff. See Autoradiography.
    binding
        host, viral, 235-44
        matrix, 240-42
    cell matrix and protoplasm, 185
    closed mitosis and, 308-09
    corticosteroid-induced alterations, 284
    morphological effects, 285-87
    crosslinking, 223-32
    DNA. See DNA.
        attachment sequences, 186ff.
            enrichment, 188, 189
            ordering and regulation, 189
        hybridization, 186ff.
        loops, 177, 301
        nucleic acid function, 293-301
        physiological association of newly-replicated with, 169, 178
        polyoma, 213-20
        viral, 214
        replication, 183-95, 293-4
    estrogen
        binding, 259-67
        isolation procedure, 41
        preparation for enzyme reactions, 32
    protein kinase
        phosphatase reactions, 31-43. See Protein Kinase, Phosphatase.

322 / Index

specific properties, 36–43
structure, 31
RNA transport, 129–33. See RNA Transport.
  binding of immunoglobulin/ATP-dependent RNP, 130. See Immunoglobulin, ATP, RNP.
  molecular level, 129ff.
  tryptophan and, 111–24. See Tryptophan.
  volume control, 315
Nuclear Envelope Reformation
  metaphase cell test population, 47
  phospholipid metabolism, 47–61. See Phospholipid Metabolism.
  HeLa S_3 cells, 47. See HeLa S_3.
Nuclear Extracts, 68. See Nuclear Binding.
  injection and preparation, 69
Nuclear Lamina. See *Drosophilia*.
  characterization, 145
  subnuclear fractions, 145
Nuclear Lamins
  function, 165
  homotypic oligomers, 164
    identification, 157ff.
    localization, 162, 163
      nuclear interior and periphery, 163ff. See Nuclear Periphery.
    structures, 267
      type II, 266ff. See Type II Estrogen.
  eucaryoles, 308
  evolution, 307–15
  function, 185ff.
    RNA-related, 185
    structural, 185ff.
  isolation medium
    G_2 phase, 207ff.
    Mg^{++}, 205
    replication forks, 204ff.
    thymidine incorporation, 208
  nuclear-reactant drugs and, 279–90
    cytotoxic, 279ff.
    drug targets, 280, 283–4
    resistance and heterogenous cell population, 287, 288

origin, 309–11
P. polycephalum, 174ff. See *P. polycephalum*.
proteins in association with HeLa, hnRNA, 238–40. See HeLa, hnRNA.
RNA processing, 293–4
splicing, 242
  pre mRNA, 243–4. See pre mRNA.
steroid hormone action, 247–55. See Steroid Hormones.
  binding and absorption, 254
  nonphysiological reaction, 255
  uniform distribution, 252
structure, 307
volume control, 315
Nuclear membrane. See Nuclear-Reactant Drugs, Proteins.
  drug targets, 280, 283–4
  change in, 289–90
  nuclear matrix and nuclear-reactant drugs, 279–90
Nuclear Morphology. See Crosslinking.
Nuclear Periphery. See Nuclear Lamins.
Nuclear Pore Complex. See *Drosophilia*, Polyanions, Proteins.
  chaotropic reagent dissection, 23ff.
  modified structure, 23
  morphology, 23
  chemical nature, 14, 15
  electrostatic interactions, 20
  envelope structure, 13–27. See Eucaryotic Cells.
  functions, 14
  isolation, 13–27
  models, 14
    biochemical, 17–19
  polyanions
    dissolution by, 20
  polydisperse polypeptides, 13–27
  transport mechanism, 1, 2
    evolution and function, 7
    selection mechanism, 7. See RNA, tRNA recognition (transient covalent bonds, protein modification, recognition signals).
  sievelike, 8
  signals, 78
  temporary assembly, 6–7
  transcription sites, 7

Nuclear Pores. See Antibodies, Lamins.
  laminas and, 163
Nuclear Protein Concentration. See
    Xenopus laevis.
  binding within, 80
  classes, 76
  cytoplasmic ratios, 80
  intercellular distribution, 76
  mechanisms of, 75-82
  n-proteins in, 75, 76. See N-Proteins.
  polypeptide quantification, 78
Nuclear Proteins
  endogenous, 64-71
    selective binding, 65
  equilibration time in polypeptides, 67
  nuclear accumulation from puncture,
      65, 66
  oocyte uptake, 63-73. See Nuclear
      Envelope and Oocyte.
Nuclear-Reactant Drugs. See RNA,
    Nuclear Matrix.
  cell death and, 281
  matrix and membrane interaction,
      279-90. See Nuclear Matrix
      and Nuclear Membrane.
  pharmacology of, 282-3
Nuclear Transport. See Nucleocyto-
    plasmic Transport.
  antibodies, 6
  continuous, 2
  energy factors, 6
  morphological examination
    elements, 2, 3
    enzymes, 2, 8
  structure movement models, 5
Nucleic Acid. See Nuclear Matrix.
Nucleocytoplasmic Transport Mecha-
    nisms, 1-10
  molecule exchange, 9

Oocytes. See Nuclear Proteins, *Xenopus
    laevis.*
  nucleus compared to erythrocyte, 136
  polypeptides of
    location, distribution, properties,
        135ff.
  Xenopus laevis
    nuclear protein uptake, 63-73
    selective process, 63

α-Polymerase. See α-Polymerase and
    DNA.

P. Polycephalum. See also Nuclear
    Matrix.
Phosphatase. See Nuclear Envelope.
Phospholipid Metabolism. See Late
    Mitotic Cells.
  analysis by rapid labeling, 50-57
  distribution, 53
  M-G$_1$ transition, 53
  classes, 52
  envelope formation and, 56-58. See
      Nuclear Envelope.
  fatty acid content and, 60-61
  inhibitors, 58
  nuclear envelope, 57-60. See
      Nuclear Envelope.
Physarum. See DNA.
Polyanions. See Nuclear Pore Complex.
  pore complexes and, 20.
Polyoma DNA Replication. See DNA,
    Nuclear Matrix.
  isolation, 214
  nuclear matrix, 213-20
  replication, 214
Polyvinylsulfate. See Nuclear
    Envelope Fractions.
Pores. See Nuclear Pores.
Pre mRNA. See Nuclear Matrix.
Proteins. See Nuclear Proteins.
  cytoplasmic. See RNA.
    RNA transport, 87
    purification of, 87
    cAMP, 88
  DNA protein. See DNA, 2M Na Cl.
  2M Na Cl resistant
    composition, 201
    mammalian cells, 199ff.
    sedimentary component, 200
  Kinase. See Nuclear Envelope.
  nonhistone
    crosslinked, 223-32. See snRNA,
        hnRNA and Histones.
    disulfide bonds
    extract, 227
    H3, 224ff.
    linking, 220. See H3 M, MA
        46,000.

RNA. See Cytoplasmic Protein, Nuclear
    Matrix and Pore Transport.
  binding
    host and viral to matrix, 235-44.
        See Nuclear Matrix.

efflux. See DNA, Mg^{2+}, NTPase
    incubation medium, 94
        C_2, 105. See C_2.
    protein release, 102. See Cytosol.
    RNA release, 94ff.
        liver sap components, 116–17
        tryptophan with, 111–24. See Tryptophan.
    envelope and matrix
        expansion and contraction, 93
        population, 93
        preparations, 92
    translocation, 274. See Ribonucleoproteins.
    transport. See ATP, Nuclear Envelope.
        agents, 89, 90
        B-γ bond, 87. See Bγ bond.
        cell-free system, 85–90
            inhibition, 86–87
            labeling, 87
        nuclear envelope, 85–90
    types
        hnRNA. See Nuclear Matrix, Proteins, Nonhistone.
            C-group proteins, 240. See C-Group
            host and viral, 235–44
            isolation of nuclear matrices, 236ff.
                steps, 236
            nuclear matrix and proteinaceous structure role, 238
            transcript processing, 238
        SnRNA. See Protein, Nonhistone.
            crosslinked with AMT to hnRNA, 228ff.
            hnRNA, 223–32
            nuclear structure maintenance, 224
            U1 & U6, 228, 230. See U1, U6.
        tRNA recognition. See Nuclear Pore Transport.
RNAase A. See *Drosophilia Melanogaster*.
RNP. See Nuclear Envelope, RNA Transport.
Ribonucleoprotein particles. See RNA.
    mechanisms and maturation, 272
    nucleocytoplasmic transport model, 271–76
    particle size, 271
    process energetics, 273–74
    ribosomes, 275–76.
    structure, 271–72

SDS-Page. See *Drosophilia Melanogaster*.
Steroid Hormones. See Nuclear Matrix.
Steroid Hormone Action
    binding sites, 247, 248, 253
    chromatin, 255. See Chromatin.
    degradation, 248
    estradiol, 249
    internal matrix structures and, 253
    peripheral lamina. See Lamina.
    salt resistant, 254. See Im Na Cl.

T Antigen. See Antigen.
Thymidine. See Nuclear Matrix.
Triton X-100. See *Drosophilia Melanogaster*.
Tryptophan. See Nuclear Envelope, RNA Efflux.
    administration, 113ff.
    amino acid limiting, 113
    cytoplasmic role, 116, 120
    intranscriptional control, 114
    nuclear RNA release envelope, 111–24
    nucleocytoplasmic translocation in RNA, 114, 115
    poly (A)-mRNA/hepatic protein, 111, 117–20
        hepatotoxic agents, 112
        hepatic polyribosomes, 112
    protein synthesis, 112, 123
    translation control, 113–114
2M Na Cl. See DNA, Proteins, Steroid Hormone Action

U1, U6. See SnRNA.
Urea Treatment. See Nuclear Envelope Fractions.

Xenopus laevis
    antibody binding, 139
    antisera, 138ff.
    components, 136
    Erythrocytes and, 135–43
    Karyoskeletal proteins, 135ff.
    nuclear protein concentration and, 76ff.
    polypeptics detected, 135, 136
    typtic peptide analysis of, 138ff.